SETON HALL UNIVERSITY
3 3073 10079481 5

D0706562

Individual Strategy and Social Structure

Individual Strategy and Social Structure

An Evolutionary Theory of Institutions

H. Peyton Young

PRINCETON UNIVERSITY PRESS
PRINCETON, NEW JERSEY

Published by Princeton University Press, 41 William Street,
Princeton, New Jersey 08540
In the United Kingdom: Princeton University Press,
Chichester, West Sussex

Library of Congress Cataloging-in-Publication Data

Young, H. Peyton, 1945–
Individual strategy and social structure: an evolutionary theory of institutions / H. Peyton Young.
p. cm.
Includes bibliographical references and index.
ISBN 0-691-02684-X (cloth: alk. paper)
1. Social institutions. 2. Institutional economics.
3. Evolutionary economics. 4. Game theory. I. Title.
HM131.Y64 1998
306—dc21 97-41419

This book has been composed in Palatino

Princeton University Press books are printed on
acid-free paper, and meet the guidelines for permanence
and durability of the Committee on Production
Guidelines for Book Longevity of the
Council on Library Resources

http://pup.princeton.edu

Printed in the United States of America

1 3 5 7 9 10 8 6 4 2

For Fernanda

[T]he problem of a rational economic order is determined precisely by the fact that the knowledge of the circumstances of which we must make use never exists in concentrated or integrated form, but solely as the dispersed bits of incomplete and frequently contradictory knowledge which all the separate individuals possess. . . .

The problem is thus in no way solved if we can show that all of the facts, *if* they were known to a single mind (as we hypothetically assume them to be given to the observing economist) would uniquely determine the solution; instead, we must show how a solution is produced by the interactions of people, each of whom possesses only partial knowledge.

—Friedrich von Hayek,
"The Use of Knowledge in Society"

CONTENTS

PREFACE

THIS BOOK is a slightly lengthened version of a series of lectures I gave in June 1995 at the Summer School in Economic Theory at the Institute for Advanced Studies of Hebrew University. The goal of the book is twofold. One is to suggest a reorientation of game theory in which players are not hyper-rational and knowledge is incomplete. In particular, I dispense with the notion that people fully understand the structure of the games they play, that they have a coherent model of others' behavior, that they can make rational calculations of infinite complexity, and that all of this is common knowledge. Instead, I postulate a world in which people base their decisions on limited data, use simple predictive models, and sometimes do unexplained or even foolish things. Over time, such simple adaptive learning processes can converge to quite complex equilibrium patterns of behavior. Indeed a surprising number of classical solution concepts in game theory can be recovered via this route.

The second goal of the book is to suggest how this framework can be applied to the study of social and economic institutions. Here, I use "institution" in its everyday sense: "an established law, custom, usage, practice, organization" (Shorter Oxford English Dictionary). I take the view that institutions emerge over time from the cumulative experience of many individuals. Once their interactions coalesce into a settled pattern of expectations and behaviors, an "institution" has come into being. The theory makes qualitative predictions about the evolutionary paths that such processes tend to follow and the diversity of institutional forms that they produce. The theory also tells us something about the welfare properties of these institutions. I illustrate these ideas through a variety of simple examples, including segregated neighborhood patterns, forms of economic contracts, terms of distributive bargaining, norms of cooperation, conventions of social deference, rules of the road, and so forth. These examples are illustrative and meant to suggest directions for future work; I do not pretend to give a definitive account of the history of any one institutional form.

Like all "new" approaches, the one presented here rests on ideas that have been around for a long time. Of particular importance is the work of Thomas Schelling, who to my knowledge was the first economist to show explicitly how micro decisions by many individuals evolve into recognizable patterns of macro behavior. If this book adds anything to his, it is to provide the analytical foundations for studying these kinds of models, and to widen their sphere of application. The second shoulder on which this book rests is the work of biologists Maynard Smith and Price. Like Schelling, they showed how game theory provides the crucial link between the micro behavior of individuals and the aggregate behavior of populations. As biologists, however, they were not motivated by the idea that individuals respond rationally to their environment. Instead, they maintained, poorly adapted individuals (animal or human) are weeded out by natural selection: those who play the game well breed faster than those who do not play as well. This yields an evolutionary selection process known as the *replicator dynamic*.

To study this selection dynamic, Maynard Smith and Price (1973) introduced a new equilibrium concept known as an *evolutionarily stable strategy* (ESS). An ESS is a frequency distribution of strategies in the population that cannot successfully be invaded by a small group of mutants. Any such distribution must be a Nash equilibrium of the underlying game, but not every Nash equilibrium is an ESS. The evolutionary perspective thus provides a novel twist to a classical solution concept in game theory, and suggests how it can be strengthened. For a more complete account of this approach, the reader is referred to the pioneering work of Maynard Smith (1982) and the superb surveys by Hofbauer and Sigmund (1988) and Weibull (1995).

This book takes a different approach, one that is closer in spirit to Schelling than to the biologists. First, my interest is in economic and social phenomena, not the behavior of mice and ants. This calls for a different class of adaptive dynamics. Second, the standard solution concept in this literature, ESS, is not sharp enough for my purposes. Indeed, it was dissatisfaction with this idea that led Dean Foster and me to formulate an alternative solution concept known as stochastic stability, which is the foundation of much that follows. Roughly speaking, an equilibrium is "stochastically stable" if it is robust against persistent random shocks, not just isolated shocks, as is assumed for ESS. This leads to a much sharper notion of equilibrium (and disequilibrium) selection, as I shall show in subsequent chapters.

The book is necessarily somewhat technical, but I have tried to keep it unencumbered by lengthy proofs, which have been relegated to the Appendix. A rudimentary acquaintance with game theory is presupposed; the required elements of dynamical systems theory and Markov processes are developed from scratch. The material should be easily accessible to graduate students and professional economists, and it is not beyond the reach of advanced undergraduates. The lectures were delivered in five sessions of about an hour and a half each. The book is longer, but not by much, so it can easily serve as a module in a traditional game theory course.

ACKNOWLEDGMENTS

THIS BOOK builds on the ideas of many others. I am particularly indebted to Dean Foster, who was a crucial partner in developing the general approach described here, and to Yuri Kaniovski, who introduced me to stochastic approximation theory and its application to urn schemes. Others whose work I draw on directly or indirectly include Brian Arthur, Robert Axtell, Ken Binmore, Larry Blume, Glenn Ellison, Joshua Epstein, Drew Fudenberg, Michihiro Kandori, Alan Kirman, David Levine, George Mailath, Richard Nelson, Georg Nöldeke, Douglass North, Rafael Rob, Larry Samuelson, Andrew Schotter, Reinhard Selten, Robert Sugden, Fernando Vega-Redondo, Jörgen Weibull, and Sydney Winter.

The exposition benefited from comments by students at the Jerusalem Summer School in Economic Theory, the Stockholm School of Economics, the University of Paris, Johns Hopkins University, and the University of Chicago. I particularly want to express my appreciation to Marco Bassetto, Joe Harrington, Josef Hofbauer, Cassey Kim, Sung Kim, Stephen Morris, Adriano Rampini, Philippe Rivière, John Rust, and Sarah Stafford, who read the manuscript with a critical eye and offered many constructive suggestions. Special thanks are due to Todd Allen, who ran the simulations reported here. Peter Dougherty's enthusiasm for the project helped bring it to fruition; Lyn Grossman did a superb job of copyediting the final product.

I am indebted to the John D. and Catherine T. MacArthur Foundation, the Pew Charitable Trusts, and The Brookings Institution for providing generous financial support. The ideas expressed herein do not necessarily reflect the views of any of these institutions.

Finally, I owe a substantial intellectual debt to Kenneth Arrow and Thomas Schelling, who have provided personal encouragement and professional inspiration over many years. Debts like this can never be properly repaid; one can only hope to pass them on to another generation.

Individual Strategy and Social Structure

Chapter 1

OVERVIEW

ECONOMIC and social institutions coordinate people's behaviors in various spheres of interaction. Markets coordinate the exchange of particular kinds of goods at specific times and places. Money coordinates trade. Language facilitates communication. Norms of etiquette coordinate how we interact socially with one another. The common law defines the bounds of acceptable behavior with respect to persons and property, and tells us what to expect when we overstep these bounds. These and many other institutions are the product, at least in part, of evolutionary forces. They are shaped by the cumulative impact of many individuals interacting with one another over long periods of time. Markets often grow up at convenient meeting places, such as a crossroads or a shady place (under the buttonwood tree). Customers come to expect particular kinds of goods to be offered there, and sellers come to meet their expectations. They also come to expect certain days and hours of operation, and particular procedures governing trade, whether posted price, haggling, or auction. These features are determined to a considerable degree by the accumulation of historical precedents, that is, by the decisions of many individuals who were concerned only with making the best trade at the moment, not with the impact of their decisions on the long-run development of that market.

A similar argument applies to economic contracts. When people rent apartments, for example, they are typically presented with a standard lease; usually the only things negotiated are price and the period of occupancy. People prefer standard contracts because they are more clearly enforceable in court than contracts that are fashioned on the spot. The accumulation of precedent makes them better defined, and hence more desirable to both parties to the transaction. But how do standard contracts become standard? The answer, evidently, is through a long period of experimentation with different forms. Eventually one form becomes standard and customary for a given type of transaction (in a given locale), not necessarily because it is optimal, but because it serves the purpose reasonably well and it is what everyone has come to expect. It is now an institution that coordinates behaviors, and to deviate from it would be costly.

A similar argument can be made for a great variety of social and economic institutions—language, codes of dress, forms of money and credit, patterns of courtship and marriage, accounting standards, rules of the road. In most cases, no one willed them into being: they are what they are due to the accumulation of precedent; they emerged from experimentation and historical accident.

Of course not all institutions can be explained in this way. Some were created by edict. In the Middle Ages, market towns were often established by royal charter. Rules of the road are enshrined in statutes. Accounting standards are regulated by official or semiofficial bodies. Languages are taught from standard dictionaries and grammar books. When we look more deeply into the matter, however, we find that these codifications were often just a way of ratifying practices that had already come into being through evolution. Furthermore, codifications and edicts do not stop evolution in its tracks: market towns come and go, accounting standards change, dictionaries and law books are always being rewritten.

The notion that economic institutions and patterns of behavior can be explained as the product or outcome of many individual decisions is scarcely a new idea in economics. It is perhaps most prominently associated with members of the Austrian school, notably Menger, von Hayek, and Schumpeter, though elements of the approach are implicit in the writings of earlier authors, including Adam Smith, David Hume, and Edmund Burke.[1]

What are the features that distinguish the "evolutionary" from the classical point of view in economics? One is the status accorded to equilibrium; the other is the status accorded to rationality. In neoclassical economics, equilibrium is the reigning paradigm. Individual strategies are assumed to be optimal given expectations, and expectations are assumed to be justified given the evidence. We, too, are interested in equilibrium, but we insist that equilibrium can be understood only within a dynamic framework that explains how it comes about (if in fact it does). Neoclassical economics describes the way the world looks once the dust has settled; we are interested in how the dust goes about settling. This is not an idle issue, since the business of settling may have considerable bearing on how things look afterwards. More important, we need to recognize that the dust never really does settle—it keeps moving about, buffeted by random currents of air. This persistent buffeting by random forces turns out to be an essential ingredient in describing how things look on average over long periods of time.

The second feature that differentiates our approach from standard ones is the degree of rationality attributed to economic agents. In neoclassical economic theory—especially game theory—agents are assumed to be hyper-rational. They know the utility functions of other agents (or the probability that other agents have these utility functions), they are fully aware of the process they are embedded in, they make optimum long-run plans based on the assumption that everyone else is making optimum long-run plans, and so forth. This is a rather extravagant and implausible model of human behavior, especially in the complex, dynamic environments that economic agents typically face. Moreover it represents a peculiar aberration from traditional ways of thinking in economics. One of the central messages of the pure theory of exchange, for example, is the ability of prices and markets to coordinate economic activity *without* assuming that agents are anything more than naive optimizers acting on limited information.

In this sense, our point of view represents a return to older traditions in economics. Agents adapt—they are not devoid of rationality—but they are not hyper-rational. They look around them, they gather information, and they act fairly sensibly on the basis of their information most of the time. In short, they are recognizably human. Even in such "low-rationality" environments, one can say a good deal about the institutions (equilibria) that emerge over time. In fact, these institutions are often precisely those that are predicted by high-rationality theories—the Nash bargaining solution, subgame perfect equilibrium, Pareto-efficient coordination equilibria, the iterated elimination of strictly dominated strategies, and so forth. In brief, evolutionary forces often *substitute* for high (and implausible) degrees of individual rationality when the adaptive process has enough time to unfold.

Let us add a little more meat to this rather skeletal outline. Recall that our general objective is to show how economic and social institutions emerge from the interactive decisions of many individuals. To talk about this idea rigorously, we need a model of how individuals interact at the micro level. This is naturally provided by a *game*, which describes the strategies available to each player and the payoffs that result when they play their strategies. Obviously the form of the game depends on the interactive situation we are trying to model. To illustrate ideas, we shall usually rely on games having a relatively simple structure, such as coordination games and bargaining games, but the theory extends to all finite-strategy games, as we show in Chapter 7.

The framework differs from traditional game theory in several crucial respects, however. First, players are not fixed, but are drawn from a large population of *potential* players. Second, the probability that individuals interact depends on exogenous factors, such as where they live, and more generally on their proximity in some suitably defined social space. Third, agents are not perfectly rational and fully informed about the world in which they live. They base their decisions on fragmentary information, they have incomplete models of the process they are engaged in, and they may not be especially forward looking. Still, they are not completely irrational: they adjust their behavior based on what they think other agents are going to do, and these expectations are generated endogenously by information about what other agents have done in the past. On the basis of these expectations, the agent takes an action, which in turn becomes a precedent that influences the behavior of future agents. This creates a feedback loop of the following type:

$$
\begin{array}{ccc}
\text{Precedents} & \longrightarrow & \text{Expectations} \\
\nwarrow & & \swarrow \\
& \text{Actions} &
\end{array}
$$

Finally, we assume that the dynamic process is buffeted by random perturbations that arise from a variety of factors, such as exogenous shocks or unpredictability in people's behavior. These shocks play a role similar to that of mutations in biology by constantly testing the viability of the current regime. Moreover, they imply that *the evolutionary dynamic never settles down completely; it is always in flux*. The novel element of the approach from a technical standpoint is to show explicitly how to analyze the long-run behavior of such processes.

To illustrate these ideas concretely, consider the following variant of Schelling's model of neighborhood segregation patterns (Schelling, 1971, 1978). There are two types of people—As and Bs—who choose where they want to live. Their utility for a given location depends on the composition of the neighborhood, that is, on the mixture of As and Bs around them. The situation is depicted in Figure 1.1, where each circle represents a location. Let us suppose that an individual is *discontent* if his two immediate neighbors are unlike himself; otherwise, he is *content*. An equilibrium configuration is one in which no two individuals want to trade places. In other words, there is no pair such

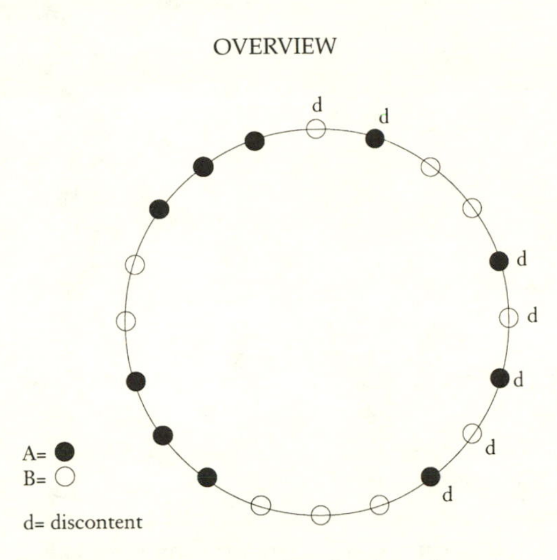

Figure 1.1. A random distribution of two types of agents on the circle.

that one (or both) is currently discontent and both would be content after they trade locations. (If only one person is discontent beforehand, we can imagine that she compensates the other to move, so that both are better off after the move than they were before.)

We claim that if there are at least two people of each type, then in equilibrium no one is discontent. To see this, suppose to the contrary that an A is surrounded by two Bs (... BAB ...). Moving clockwise around the circle, let B* be the last B-type in the string of Bs who follow this A, and let A* be the person who follows B*:

$$\ldots BAB \ldots BB^*A^* \ldots .$$

Since there are at least two agents of each type, we can be sure that A* differs from the original A. But then the original discontented A could switch with B* (who is content), and both would be content afterwards. Thus we see that the equilibrium configurations consist of those arrangements in which everyone lives next to at least one person of their own type. No one is "isolated."

In general there are many types of equilibrium configurations. Some consist of small enclaves of As and Bs scattered around the landscape; others are completely segregated in the sense that all As live on one side of the circle, and all Bs on the other side (see Figure 1.2).

So much for the equilibrium analysis. What happens when the process begins in an out-of-equilibrium situation—will equilibrium even-

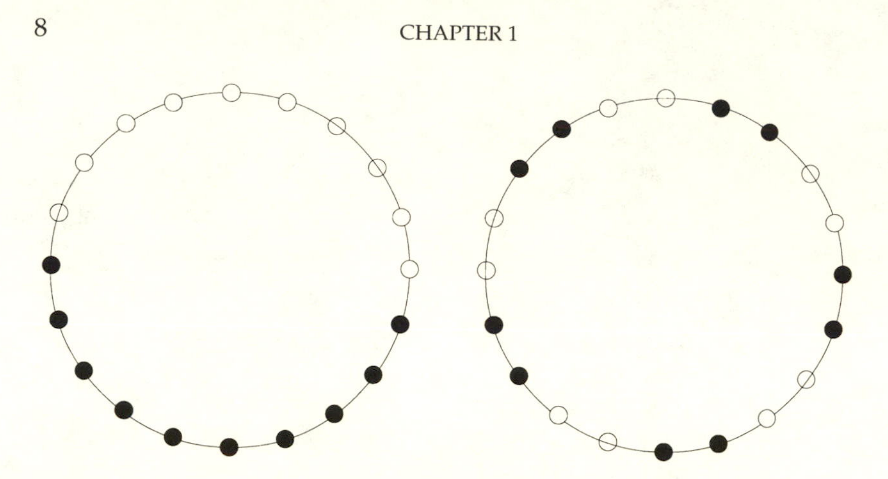

Figure 1.2. Two equilibrium configurations: one segregated, the other integrated.

tually be reached? Consider the following adjustment dynamic. Time is broken into discrete periods. In each period, two people meet at random and have a conversation about their respective neighborhoods. If they find that they can trade places to advantage, they do so. The reader may check that from any initial state, there always exists some sequence of advantageous trades that leads to an equilibrium. Since the number of states is finite and the probability of following such a path is positive, the process will eventually find its way to some equilibrium with probability one. Figure 1.3 illustrates a series of adjustments of this type. Note that the adjustment process is not fully predictable: where it ends up depends on where it starts and on the order in which people happened to trade. One can therefore speak of the probability of reaching various outcomes from some initial state, without knowing which one will in fact materialize. Such processes are sometimes said to be "path-dependent."[2] Of course, any nontrivial stochastic process is path dependent in the sense that different paths will be followed depending on the outcome of chance events. A more telling definition of path dependency is that, with positive probability, the process follows paths that have different long-run characteristics (the process is "nonergodic"). The location model described above is path dependent in this sense, because the paths end up in different equilibrium configurations.

A further complication arises from the fact that people do not always act the way our models say they do. Suppose, for example, that there is a small probability that a pair of individuals who could gain from

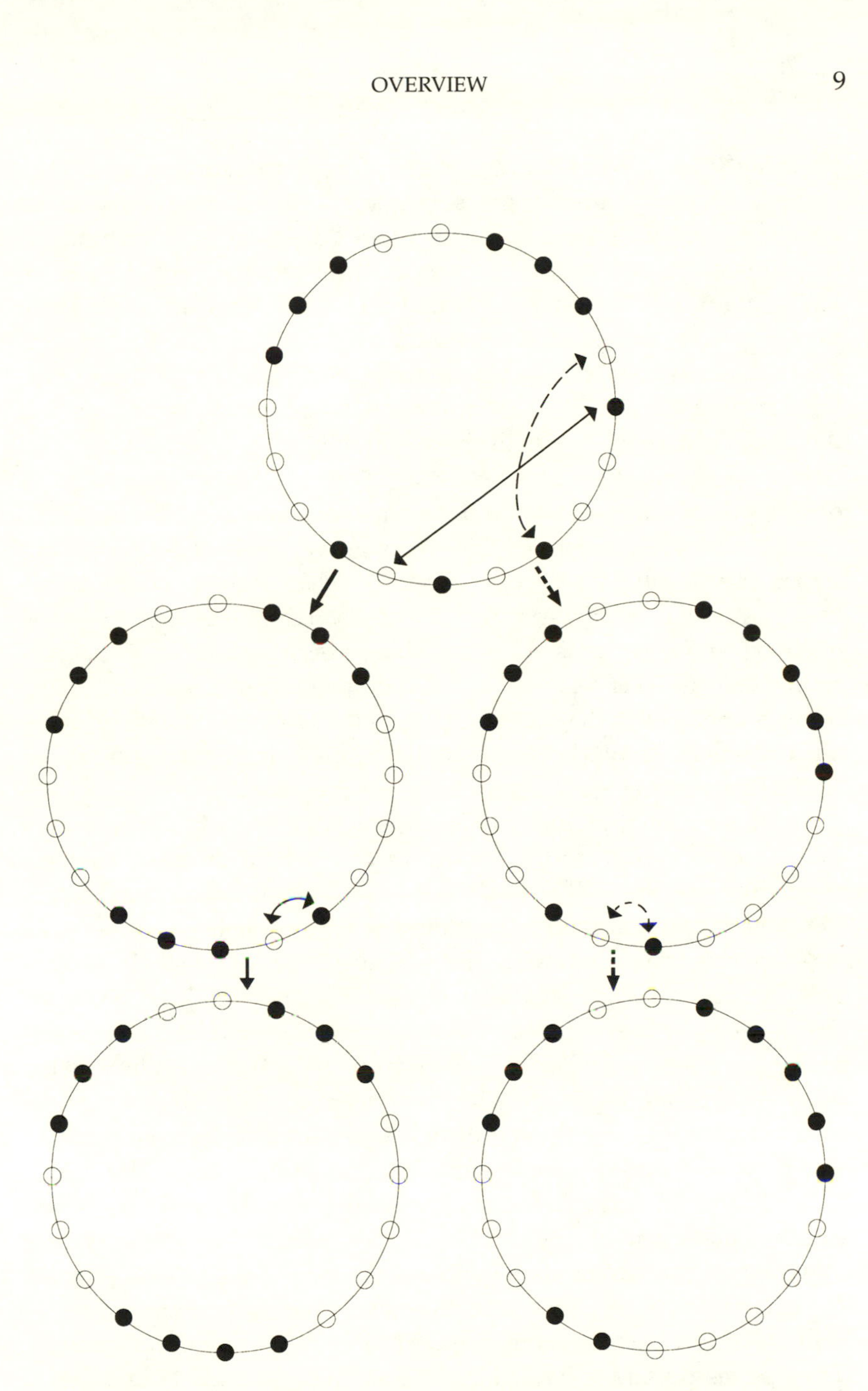

Figure 1.3. Alternative evolutionary paths from same starting point.

trading places will nevertheless fail to do so; similarly, there is a small probability that a pair who cannot gain from trading places will trade anyway. This assumption expresses the simple fact that we do not know all the reasons why people act the way they do. Our behavioral model is incomplete, and actions need to be modeled as random variables. These kinds of uncertainties have important implications for the behavior of the process. In particular, the process is *ergodic*, that is, its long-run average behavior is essentially independent of the path taken; furthermore it is independent of the initial conditions.[3] Using techniques that we shall develop in Chapter 3, we shall show that the process spends almost all of its time in one of the completely segregated states where all As live on one side of the circle and all Bs on the other side.

Note that nobody intended this outcome; it arises because people optimize locally and do not worry about the effect of their actions on the long-run properties of the system. Note also that the process keeps on evolving—it never "ends up" somewhere—because the residential pattern of As and Bs is constantly shifting: sometimes As will live in the north, sometimes in the south; segregated neighborhoods will eventually become integrated, later they will become segregated again, and so forth. This represents a fairly accurate picture of reality.

While we cannot predict the dynamic path that such a process will follow, we can estimate the *probability* with which different kinds of residential patterns will be observed over the long run. In the present example, the answer is all too familiar: when people have a preference for some neighbors who are similar, and everyone is left to their own devices, completely segregated neighborhoods are more likely to emerge than any other pattern. In fact, this remains true even if everyone *prefers* to live in a mixed neighborhood (where one neighbor is similar and the other is different), than to live in a neighborhood surrounded by their own kind. In this case the segregated states may be far from optimal, but in the long run they are the most likely states.

The analytical techniques that allow us to prove this result are developed in subsequent chapters. There is, however, an important concept at work here that can be stated without the formal apparatus: when an evolutionary process is subjected to small, persistent stochastic shocks, some states occur much more frequently than others over the long run. These states are said to be *stochastically stable* (Foster and Young, 1990). Later we shall show how to compute the stochastically stable states explicitly. What bears emphasizing here is that stochastic stability is a considerably sharper (and more general) criterion of equilibrium selection than such established notions in the literature as "evolutionary stable strategy" and "risk-dominant equilibrium," though it is related to them in special cases, as we shall see in subsequent chapters.

To illustrate the kinds of questions we can address using this concept, let us briefly consider several further stylized examples. One economic institution that clearly has an evolutionary flavor is the choice of medium of exchange.[4] History reveals the great variety of goods that societies have adopted as money: some used gold or silver; some, copper or bronze; others used beads; still others favored cattle. In the early stages of economic development, we can conceive of the choice of currency as growing out of individual decisions that gradually converge on some norm. Once enough people in a society have adopted a particular currency, everyone else wants to adopt it, too.

At the most basic level, this kind of decision problem can be modeled as a coordination game. Suppose that there are two choices of currency: gold or silver. At the beginning of a period, each person must decide which currency to carry (we assume that carrying both is too costly). During the period, each person meets various other people in the society at random, and they can trade only if they are both carrying the same currency. Thus the decision problem at the beginning of the period is to choose the currency that one thinks will be chosen by a majority of the others.

Schematically, we can model this situation as follows. Let p^t be the proportion in the population choosing gold at time t, and let $1 - p^t$ be the proportion choosing silver. In period $t+1$, some people reconsider what they are doing (or they die and are replaced by people who must make a new decision). For the sake of simplicity, suppose that exactly one person, drawn at random from the general population, reconsiders during each period. Assume that the properties of gold and silver make them equally desirable as currencies. (We shall relax this assumption in a moment.) Then our decisionmaker chooses gold if $p^t > .5$ and chooses silver if $p^t < .5$. If $p^t = .5$, we can assume that the decisionmaker continues to do whatever he was previously doing because of inertia.[5] All of this happens with high probability, say, $1-\varepsilon$. But with probability $\varepsilon > 0$ a person chooses gold or silver at random, that is, for reasons external to the model.

Qualitatively, this process evolves in the following manner. After an initial shakeout, the process converges quite rapidly to a situation in which most people are carrying the same currency—say, gold. This norm will very likely stay in place for a considerable period of time.

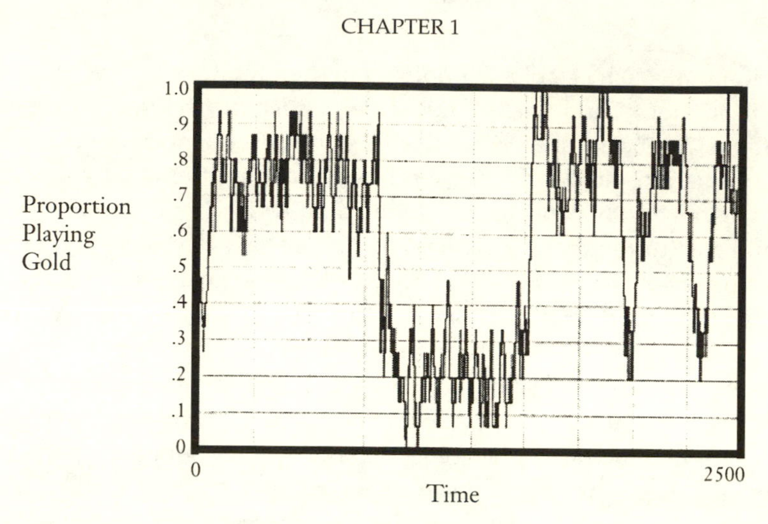

Figure 1.4. The currency game with equal payoffs, population size 10, and $\varepsilon = .5$.

Eventually, however, an accumulation of random shocks will "tip" the process into the silver norm. These tipping incidents are infrequent compared to the periods in which one or the other norm is in place (assuming ε is small). Moreover, once a tipping incident occurs, the process will tend to adjust quite rapidly to the new norm. This pattern—long periods of stasis punctuated by sudden changes of regime—is known in biology as the *punctuated equilibrium effect*. The evolutionary model we have described above predicts that a similar phenomenon characterizes shifts in economic and social norms. Figure 1.4 illustrates this phenomenon for the currency game when the two currencies have equal payoffs.

Now let us ask what happens when one currency is inherently a little better than the other. Suppose that gold is somewhat preferred because it does not tarnish as easily as silver. Then the decision problem at the individual level is to choose gold if $p^t > \gamma$, and to choose silver if $p^t < \gamma$, where γ is some fraction less than .5 but larger than 0. Now the process follows a path that looks like Figure 1.5. Over the long run, there is a bias toward gold; that is, at any given time, the society is more likely to have adopted the gold standard than to have adopted the silver standard. This is not surprising. What is perhaps surprising is that the bias becomes larger the smaller the random perturbations are. Figures 1.6 and 1.7 show characteristic sample paths for $\varepsilon = .10$ and $\varepsilon = .05$. It is clear that the smaller the value of ε, the more likely it is that the process is in a gold standard phase, and the longer are the

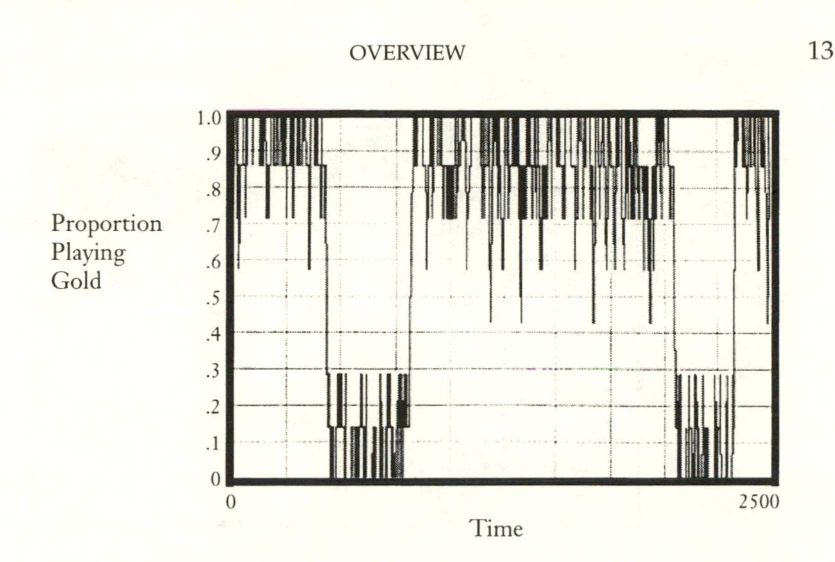

Figure 1.5. The currency game with asymmetric payoffs (gold = 3, silver = 2), population size 10, and $\varepsilon = .5$.

periods between shifts of regime. This can be verified analytically by computing the long-run distribution explicitly as a function of ε, γ, and the population size, as we show in Section 4.5.

A similar argument can be applied to a variety of other phenomena. Consider, for example, a competition between new technologies that have a networking effect. A contemporary example is personal computers: if most people own IBMs, it is advantageous to own an IBM; if most people own Macs, it is more desirable to own a Mac. The reason is that the more popular a given model is, the more software will be created for it, and the easier it will be to share programs with others.[6]

This situation can be modeled dynamically as follows. Imagine that there are m potential customers for two competing technologies. For the sake of concreteness, we shall stick with the case of personal computers. In each period, one person decides to buy a new computer. In the start-up phase, he may be a first-time buyer; in later phases, he will be replacing a computer he already owns. Suppose first that IBMs and Macs are equally desirable from the standpoint of operation and cost; the only thing that differentiates them is their current popularity. As in the currency game, there is a high probability that the customer will choose the more popular computer and a low probability that he will choose the other. Or, to be more realistic, we can suppose that the customer samples from the current population of users in order to estimate the one that is most popular, and then makes her choice. (This "par-

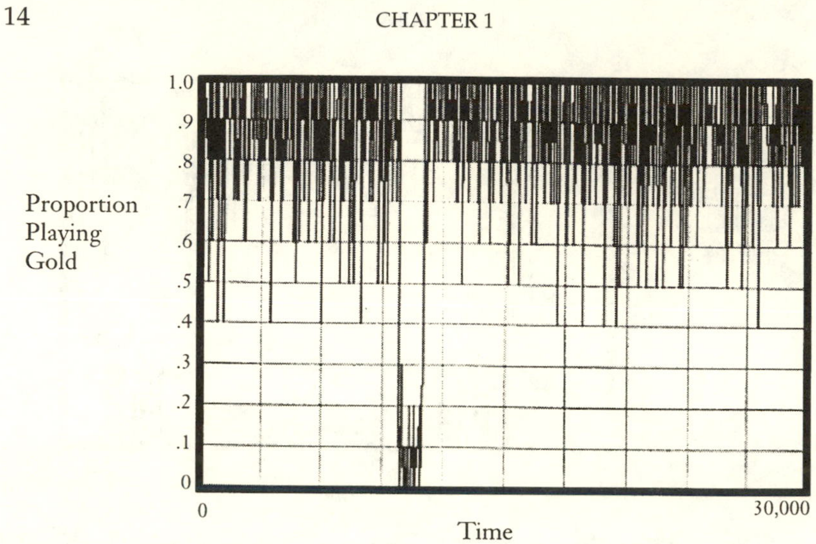

Figure 1.6. The currency game with asymmetric payoffs, population size 10, and $\varepsilon = .10$.

tial information" model has long-run properties very similar to those of the full information model, as we shall show in later chapters.) Just as in the currency game, the process oscillates between periods in which first one technology and then the other is dominant. However, if one technology is better (even a little better) than the other, the model predicts that over the long run the process will find its way to the superior technology, which will tend to stay in place for longer periods of time than will the inferior one. In our terminology, the superior technology is stochastically stable.

This result must be interpreted with some caution, however, because the long run in such a model can be very long indeed compared to the rate of technological change. By the time the process finds its way to the superior technology, the nature of the two technologies may have changed entirely. As Arthur (1989) has argued, what matters from a practical standpoint is who seizes the largest share of the market initially, for this confers an advantage on the leader that can be very difficult to overcome in the short or intermediate run. To put it another way, if an inherently inferior technology gets a head start on an inherently superior one as a result of chance events, and if there are strong networking effects, then the inferior technology is likely to hold onto its lead for a long period of time before stochastic forces displace it in favor of the superior technology.[7]

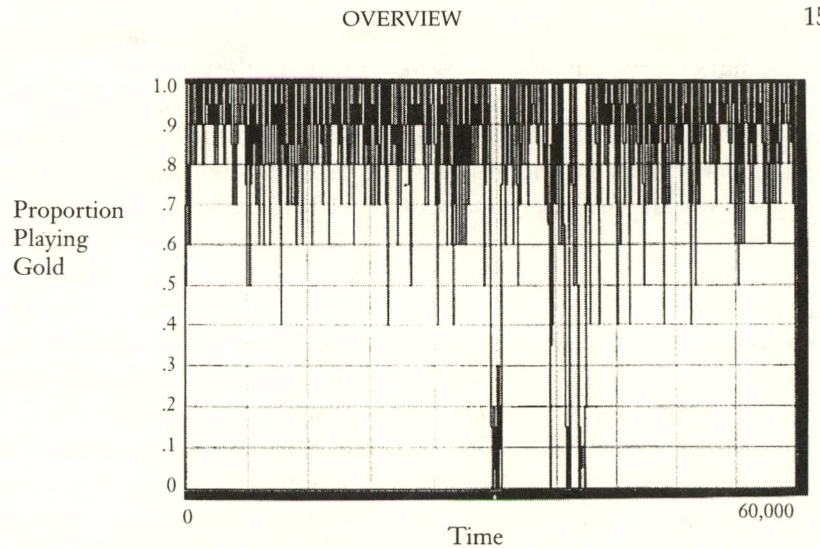

Figure 1.7. The currency game with asymmetric payoffs, population size 10, and $\varepsilon = .05$.

Thus, from a short-run perspective, a key property of the system is its *inertia*, that is, the expected waiting time until the process tips from the less favorable to the more favorable regime. While this depends in part on the comparative superiority of one technology over the other, it also depends on (i) the size of the customer base, (ii) the amount of information on which customers base their decisions, (iii) the magnitude of the stochastic perturbations, and (iv) the extent to which customers gather information locally or globally. When people have a large information base and they interact globally, the inertia can be enormous: once an inferior technology gets a jump on a superior one, it can take almost forever for evolutionary forces to dislodge it. On the other hand, when people base their decisions on relatively little information, and they interact mainly with small groups of neighbors, the process can find its way to the optimum technology relatively quickly. Note, however, that low inertia is a two-edged sword: while it reduces the waiting time to get to the optimum outcome, the optimum does not remain in place as long before it risks being displaced (by further shocks) in favor of a suboptimal outcome.

This discussion raises an important issue in evolutionary models, namely, the time scale in which events unfold. In the processes we shall study, time is measured in discrete periods that correspond to distinct events. For example, each interaction between a pair of individuals might mark the beginning of a new period. When the population is large and people interact often, thousands or even millions of such events might be compressed within a short period of real time, such as an hour or a day. It is therefore not meaningful to make statements about short- versus long-run phenomena without a metric for translating event time into real time.

An implicit assumption in these models is that some parameters are changing much more slowly than others, so that the former can be viewed as fixed compared to the latter. When we model strategic behavior in a game, for example, we usually assume that the payoff structure remains fixed while the expectations of the players change. This assumption makes more sense in some cases than in others. In the competition between computer technologies, for example, payoffs may be changing so quickly that they need to be viewed as a dynamic element of the system. In other situations, the game may evolve very slowly. Consider the problem of which side of the road to drive on. At the micro level, this can be viewed as a game between two oncoming vehicles: both drivers want to coordinate on the same convention—left side or right side—in order to avoid an accident. Whether the game is played between horse-drawn carriages or high-speed automobiles, it is essentially a coordination game whose payoff structure does not change much over time. (Of course, the absolute payoffs change—the disutility of an accident increases as vehicles become faster—but all that matters is that the two competing conventions have equal payoffs.)

Rules of the road would therefore seem to be a good vehicle for studying the long-run properties of convention formation. Moreover, the history of left-right driving conventions in Europe exhibits the qualitative patterns that an evolutionary model would predict.[8] In the early stages, when there was relatively little traffic on the roads and its range was limited, conventions grew up locally: a city or province would have one convention, while a few miles down the road another jurisdiction would have the opposite one. As use of the roads increased and people traveled further afield, these local rules tended to congeal first into regional and then into national norms, though for the most part these norms were not codified as traffic laws until well into the nineteenth century. In areas with highly fragmented jurisdictions, the congealing process took longer, as an evolutionary model would predict. Italy, for example, was characterized by highly localized left-right driving rules until well into the twentieth century.

Once conventions become established at the national level, the interactions are between countries, who are influenced by their neighbors: if enough of them follow the same convention, it pays to follow suit. Over time, we would expect a single convention to sweep across the board. While this intuition is essentially correct, it ignores the effect of idiosyncratic shocks, which can displace one convention in favor of another. Remarkably, just such a shock occurred in the history of European driving: the French Revolution. Up to that time, it was customary for carriages in France as well as in many other parts of Europe to keep to the left when passing. This meant that pedestrians often walked on the right to face the oncoming traffic. Keeping to the left was therefore associated with the privileged classes, while keeping to the right was considered more "democratic." Following the French Revolution, the convention was changed for symbolic reasons. Subsequently Napoleon adopted the new custom for his armies, and it spread to some of the countries he occupied.

From this point forward, one can see a gradual but steady shift—moving more or less from west to east—in favor of the right-hand rule. For example, Portugal, whose only border was with right-driving Spain, converted after World War I. Austria switched province by province, beginning with Vorarlberg and Tyrol in the west and ending with Vienna in the east, which held out until the Anschluss with Germany in 1938. Hungary and Czechoslovakia also converted under duress at about this time. The last continental European country to change from left to right was Sweden in 1967. Thus we see a dynamic response to an exogenous shock (the French Revolution) that played out over the course of almost two hundred years.

Of course, whether people drive on the left or the right is not particularly consequential in terms of social welfare. What matters is that society have an established convention in which expectations and behaviors are in equilibrium. It should be borne in mind, however, that some equilibria may be quite undesirable from the standpoint of social welfare. Indeed, some games are perverse in the sense that *everyone* does poorly in equilibrium. (Prisoner's dilemma is a prominent example.) Evolutionary arguments do not negate this problem; they can explain how bad equilibria come about, but they do not eliminate them.

Consider, for example, a situation in which two individuals cooperate to produce a joint product. They can either work hard or shirk. If both work, their output is high; if both shirk, their output is low. If one works and the other shirks, the output is the same as if both had shirked, but the one who worked put in more effort (in vain) and hence is less well off than the one who shirked.[9] Assuming equal division of the spoils, we have a game with the following payoff structure (the specific numbers are unimportant):

	Work	Shirk
Work	10, 10	0, 7
Shirk	7, 0	7, 7

If you expect your partner to work, it pays to work too; if you expect your partner to shirk, it is better to shirk. Thus two different norms can emerge: work or shirk. It turns out that in an evolutionary model of the type described above, shirking is stochastically stable: it is more likely to be the norm at any given time than working. The intuitive reason is that there is always some random variation in people's strategies, and hence some uncertainty about what one's partner will do. In such an environment, shirking is a safer strategy than working. Suppose, for example, that one believes that there is a greater than 30 percent chance that one's partner will shirk. Then it is best to shirk, too. On the other hand, one must believe that there is a greater than 70 percent chance that one's partner will work to make one want to work. In the terminology of Harsanyi and Selten (1988) the shirk norm is *risk dominant*.

But why would one believe that there is a 30 percent chance that one's partner is going to shirk when almost everyone is currently working? The answer has to do with variability in the population. Even if most people currently work hard, there will almost always be a few people who for some reason or other choose to shirk. Now suppose that someone outside this group of shirkers happens to interact with them for several periods. (This is particularly likely to happen if the shirkers are concentrated in some neighborhood.) This person will come to believe (given his or her limited information) that a sizable fraction of people shirk, which will induce him or her to shirk, too. This action will then be noticed by other people, which will further reinforce the idea that people are shirking. Shirking spreads by contagion. Of course, the process can also go in the reverse direction, whereby the work norm spreads by contagion. The point about the 30 percent–70 percent comparison is that it takes longer for accumulated variations to reach the 70 percent threshold than to reach the 30 percent threshold, so the waiting time to get from shirk to work is longer than the other way around.

In general, the inertia of the system—the waiting time needed to tip from one norm to another—depends in a fairly complex way on the size of the population, the extent to which people interact with their neighbors or with those far away, the amount of the information they gather, and so forth. This issue is explored in more detail in Chapters 4–6, where we show that under a wide variety of conditions, the risk-dominant norm in a 2×2 game is stochastically stable. Over the long run, it has the evolutionary advantage whether or not the outcome is socially optimal. However, while stochastic stability and risk dominance usually coincide in 2×2 interactions, they are not the same thing in general, as we show in Chapter 7.

Stochastic evolutionary models are able to make quantitative predictions about the evolution of norms and institutions (sometimes surprisingly sharp ones), but their greatest significance lies in their qualitative properties. They have a different "look" and "feel" than equilibrium models. These qualitative features are sufficiently distinctive, in fact, that there is a reasonable prospect that they could be tested against empirical data. Although this task is well beyond the scope of the present book, we can suggest several features that, in principle, are testable.

To illustrate, consider a collection of distinct societies whose members do not interact with one another. Over time, each will develop distinctive institutions to cope with various types of economic and social coordination (forms of contracts, work norms, conventions of social behavior, and the like). We can think of these institutions as particular equilibrium outcomes in a game that has many (potential) equilibria. At any given time, a given society is likely to be "near equilibrium" in the sense that almost everybody follows the behavioral pattern that is expected of them and almost everybody wants to follow this behavior given the behavior they expect of others. Note that we say *almost* instead of *all*: there will inevitably be some misfits and nonconformists who do not follow established patterns. Moreover, according to the theory, these mutant types play a crucial role in promoting long-run change: occasionally they become numerous enough to tip society from one near-equilibrium to another. In particular, different societies may be near different equilibria at any given point in time due to the vagaries of history. This fact has two general implications. On the one hand, it says that two people in similar roles are more likely to exhibit similar behaviors if they come from the same society than if they come from different societies, assuming all other explanatory variables are held constant. This is the *local conformity* effect. On the other hand, because the same process of adaptation is operating simultaneously in all of the societies, the frequency distribution of institutional forms will be fairly stable and predictable over the long run. In particular, the theory predicts that some institutions are inherently more *stable* or *durable* than others in the presence of stochastic shocks. Once established, they tend to persist for longer periods of time. Over the long run, these institutions will occur with higher frequency among the various villages. When the stochastic shocks are small, the *mode* of this frequency distribution will tend to be close to the stochastically stable institutions predicted by theory.

These two effects can in principle be identified from cross-sectional data. A third qualitative prediction, however, concerns the look of the evolutionary path and will only be revealed by time series data. The process will tend to exhibit long periods of stasis in which it is close to some equilibrium, punctuated by relatively brief periods in which the equilibrium shifts in response to stochastic shocks. We call this the *punctuated equilibrium* effect.[10] It is a well-known feature of residential segregation patterns, but also occurs in other contexts and on different time scales. The spread of right-side driving in continental Europe appears to be one example where the tipping process took two centuries to run its course. Residential patterns, by contrast, often tip in a few years.

Organization of the Book

The book's argument is organized as follows. Chapter 2 discusses various models of adaptive behavior at the individual level. These include replicator dynamics, reinforcement models from psychology, imitation, and best-reply dynamics. For the sake of expositional clarity, we choose to focus on best-reply models, although a similar analysis could be carried out for other classes of adaptive rules. This choice is also governed by the fact that best-reply models have a firm foundation in the theory of individual choice. We can therefore address a variety of issues that are beyond the scope of the other frameworks, including the effects of risk aversion and amount of information on the evolutionary selection process.

The adaptive model is an elaboration of the feedback loop described above. Players develop expectations about others' behavior based on precedent—on information about what other people have done in the past. This information is typically fragmentary and incomplete, because a given person will generally know only a small proportion of the relevant precedents, which he learns through his social network. Furthermore memory is bounded: players do not know (or perhaps do not care) about things that happened long ago; only recent events matter. On the basis of her information, a player forms a simple statistical model of how others are likely to behave. Usually she chooses a best reply given these expectations, but sometimes she makes arbitrary or unexplained choices. This simple model of adaptive behavior forms the basis of all the subsequent analysis.

In Chapter 3 we develop a conceptual framework for studying stochastic models of learning and adaptation more generally. The chapter opens with a discussion of asymptotic stability in deterministic dynamical systems, which is the standard concept in much of the evolutionary games literature. It is argued that this concept is not satisfactory as a predictor of long-run behavior when the system is subjected to persistent stochastic shocks (as almost all such systems are). The key concept of *stochastic stability* is introduced (Foster and Young, 1990). Loosely speaking, the stochastically stable states of an ergodic stochastic process are those states that occur with nonnegligible probability when the size of the stochastic perturbations is arbitrarily small. We develop a general framework for computing the stochastically stable states of any process that is Markovian, has stationary transition probabilities, and operates on a finite state space. The approach is illustrated with the neighborhood segregation model.

Chapter 4 applies this technique to the study of two-person coordination games in which each player has just two strategies. In this case, the stochastically stable outcome coincides with Harsanyi and Selten's concept of risk dominance. We explore the implications of this result for a variety of examples, including the currency game, the work-shirk game, and games of social etiquette. We then show how to compute the long-run distribution explicitly as a function of the population size and the error rate. This analysis shows that the selective bias in favor of the risk-dominant equilibrium is quite marked even when the stochastic perturbations are large, provided the population of players is also large.

In Chapter 5 we consider various refinements and embellishments of the basic learning model. First, we examine the effects of having more or less information, that is, whether having a large or a small information base is advantageous in the long run. It turns out that the answer is complex, and that having more information is an advantage in some kinds of games, but not in others. Then we show how to analyze situations where the populations are heterogeneous with respect to the individual payoff functions and the amount of information the players have. Next, we investigate the sensitivity of the selection results to different ways of modeling stochastic perturbations. In particular, we consider the case where players deviate from best reply at different rates. We also consider the possibility that they choose a nonbest reply with a probability that decreases as the prospective loss of payoff increases (Blume, 1993). Under either modification, the stochastically stable outcome in a symmetric 2×2 coordination game remains the risk-dominant equilibrium, just as in the case of uniform random errors.

Finally, we examine the situation where memory is not bounded, and players attach equal importance to all precedents no matter how dated they are. This process exhibits substantially different long-run behavior than does the process with bounded memory. In particular it is not ergodic: the process converges to an equilibrium (or near-equilibrium) regime with a probability that depends on the initial state. We argue, however, that this result is of mainly theoretical interest, since in practice past actions do *not* carry the same weight as recent ones.

Chapter 6 analyzes situations where players are located in a geographic or social space, and interact only with their "neighbors."[11] We posit a perturbed best-reply process in which each player chooses an action with a probability that decreases exponentially the lower its expected payoff is, given what his neighbors are doing. This formulation allows us to represent the long-run probability of different states as a Gibbs distribution. Once again, we find that the stochastically stable outcome in a symmetric 2×2 game occurs when everyone plays the risk-dominant equilibrium. This framework is also convenient for studying the *inertia* of the system—how long it takes (in expectation) for the process to reach the stochastically stable outcome from an arbitrary initial state. Extending previous work of Ellison (1993), we show that the inertia of a local interaction system can be dramatically lower than the inertia of a system in which everyone interacts with everyone else. In particular, if everyone interacts with a sufficiently small, close-knit group, the inertia of the process is bounded above independently of the total population size.

In Chapter 7, we extend the analysis to arbitrary finite n-person games. Here the key concept for studying the long-run selection process is sets of strategies that are closed under best replies. Specifically, a *curb set* is a Cartesian product set of strategies that includes all best replies to all possible probability mixtures over strategies in the set (Basu and Weibull, 1991). A *minimal curb set* is one that contains no smaller curb set. Building on work of Hurkens (1995), we show that in a generic n-person game, adaptive learning selects a minimal curb set; that is, in almost all n-person games there is a unique minimal curb set whose strategies are stochastically stable. We describe the general method for computing this set. In some classes of games the minimal curb sets correspond one to one with the strict Nash equilibria, in which case the learning model yields a theory of equilibrium selection. Moreover, there are important cases where the equilibria selected in these low-rationality environments are the same as those that emerge in the high-rationality world of classical theory.

Chapter 8 considers one such example, the Nash bargaining solution. In the traditional noncooperative model of bargaining, two players take turns making offers about how to divide a pie between themselves (Stahl, 1972; Rubinstein, 1982). If a player refuses an offer, the other player makes a counteroffer. If a player accepts an offer, the game ends. Assuming there is a small probability that the bargaining will "break down" after each refusal, the subgame perfect equilibrium outcome of this game is close to the division that maximizes the product of the players' utilities (the Nash bargaining solution). Note that this argument relies on the assumption that both players understand the structure of the game, that their utility functions are common knowledge, and that their rationality is common knowledge.

The evolutionary model dispenses with all three of these assumptions. As above, we assume that agents in the current period have expectations that are shaped by the precedents they know about from earlier periods. Each player normally demands an amount that has the highest expected payoff given his belief about how much the other will demand, though sometimes the demands are idiosyncratic. If the bargainers within each population have the same degree of risk aversion and the same amount of information, the stochastically stable outcome is close to the Nash bargaining solution. We therefore obtain a classical solution concept without extravagant assumptions about the agents' level of rationality or the degree of common knowledge. Moreover, when agents are heterogeneous in their characteristics, the model yields a novel generalization of the Nash solution.

A distributive bargain may be viewed as a primitive contract between two parties about dividing a pie. Chapter 9 extends the analysis to the evolution of contracts more generally. A *contract* expresses the terms that govern the relations between people. Its terms may be quite explicit, as in a bank loan, or almost entirely implicit, as in a marriage. Whether implicit or explicit, there is a tendency for contracts to follow standard formats. It is a commonplace observation, however, that what is standard in one society may not be standard in another. This fact suggests that evolutionary forces are at work in determining what types of contracts people *consider* to be standard.

To get a handle on this issue, we model the bargaining process as a pure coordination game. Each party demands certain terms in the contract. They enter into a contractual relationship if and only if their demands are consistent. Each player's expectation about what the other will accept is shaped by precedent, that is, by the terms that other people have agreed to in similar situations. This feedback effect causes society to converge toward "standard contracts"—terms that are normal and customary in a given kind of relationship. When stochastic perturbations are introduced, the model makes predictions about the welfare properties of contracts that are most likely to be observed in the long run. In particular, it says that such contracts will tend to be *efficient*: no other terms offer higher expected payoffs for both sides. Furthermore, they will tend to be *fair* in the sense that the expected payoffs are more or less centrally located within the payoff opportunity set. While similar predictions can be obtained in a world where agents employ rational principles of contract selection, no such assumption is made here. Rather, the outcome emerges (without anyone intending it) from the interaction of many myopic agents, each of whom is concerned only with maximizing his own welfare.

Chapter 2

LEARNING

WHENEVER two people approach a doorway, a coordination problem presents itself: who shall give way to the other? Suppose that they differ in some recognizable way—say, one is a man and the other is a woman—and they condition their behavior on who they are. If both play "After you, Alphonse," they stand outside the door wasting time; if both barge ahead, they collide. This interaction can be modeled as a game. Each player has two options—to yield or not to yield—and each must make a decision based on his or her expectations about what the other player is going to do. Each pair of actions produces an outcome, and each outcome yields a payoff for each player. Suppose that the payoffs are as follows (the reason for $\sqrt{2}$ will become clear in due course):

The Etiquette Game

		Women: Not Yield	Women: Yield
Men	Yield	$1, \sqrt{2}$	$0, 0$
	Not Yield	$0, 0$	$\sqrt{2}, 1$

The first number in each pair is the payoff to the row players ("Men"); the second number is the payoff to the column players ("Women"). The precise meaning of these payoffs depends on the behavioral model we have in mind. Here and in most of what follows, we shall adopt the standard interpretation that the payoffs represent the players' utilities, which satisfy the von Neumann–Morgenstern axioms. In particular, given the probabilities that a player assigns to the other players' intended actions, he or she will choose an action that maximizes his or her expected payoff. Under this interpretation, the game has two pure Nash equilibria: in one, the men defer to the women; in the other, the women defer to the men. There is also a mixed Nash equilibrium in which each side yields with probability $\sqrt{2} - 1$ and does not yield with probability $2 - \sqrt{2}$. The expected payoff from the mixed equilibrium is

$2 - \sqrt{2} = .586$, which is strictly less (for both players) than the payoffs in either of the pure Nash equilibria. Thus the pure equilibria are socially efficient, but the two sides prefer different equilibria. Any 2×2 coordination game with this general structure (though not necessarily these payoffs) is called the Battle of the Sexes.

As a second example, consider the following technology adoption game, which is much like the personal computer example discussed in Chapter 1. Two types of typewriter keyboards are available on the market—QWERTY (Q) and DVORAK (D). Each secretary must decide which keyboard to learn to use, and each employer must decide which type of keyboard to provide in the office. Let us assume for simplicity that each employer buys just one kind of typewriter and that each secretary is proficient on only one type of keyboard. Let us also suppose, for the sake of argument, that keyboard D is a bit more efficient, so if employers and secretaries coordinate on D, the payoffs are slightly higher than if they coordinate on Q:[1]

The Typewriter Game

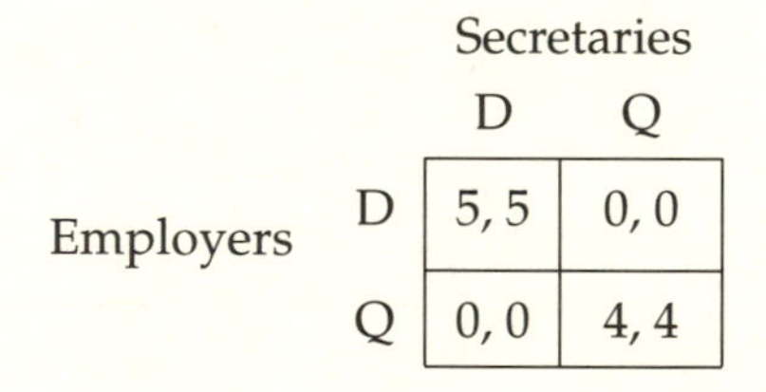

		Secretaries	
		D	Q
Employers	D	5, 5	0, 0
	Q	0, 0	4, 4

There are several important differences between this example and the preceding one. First, the structure is different: while there are two pure equilibria and one mixed equilibrium, in this case only the equilibrium (D, D) is socially efficient. Nevertheless, society can become trapped in the low-level equilibrium (Q, Q) if everyone expects everyone else to adopt Q. Second, there is a difference in the frequency with which each individual plays the game. People go through lots of doors in their lifetime, but secretaries do not learn new methods of typing very often, and employers only occasionally invest in new technology. Thus, in the first example, there is a lot of "learning by doing" that is possible at the individual level, whereas in the second example it is too costly and time-consuming for individuals to learn much from personal experience. Indeed, in both examples, it is likely that people learn partly (perhaps even mainly) by observing the actions of others. Rules of deference are learned by going through doors, and also by watching others go through doors. Secretaries and employers learn which keyboards are most popular by asking around. In short, information from previous plays of the game shapes the expectations of those who are going to play the game in the future. Our interest is in the aggregate patterns of behavior that emerge in these types of learning environments, where many dispersed agents are making myopic decisions based on fragmentary, hearsay information.

2.1 Varieties of Learning Behavior

Our first task is to specify how individuals adapt their behavior in response to others' behavior in an interactive decisionmaking environment. This is a key building block in the overall structure, yet the truth is that we do not yet have adequate data (let alone an accepted theory) about how individuals actually make such decisions. We are therefore reduced to making plausible assumptions based on everyday observation and bits and pieces of experimental evidence. Among the adaptive mechanisms that have been discussed in the learning literature are the following.

1. *Natural selection.* People who use high-payoff strategies are at a reproductive advantage compared to people who use low-payoff strategies; hence the latter decrease in frequency in the population over time. The standard model of this situation is the replicator dynamic, in which the rate of growth of a strategy in the population is assumed to be a linear function of its payoff relative to the average payoff. In this context, payoffs refer to reproductive success rates, not to individuals' preferences over outcomes.[2]

2. *Imitation.* People copy the behavior of others, especially behavior that is popular or appears to yield high payoffs. Imitation may be driven purely by the behavior's popularity (copy the first person you see), or there may be a relationship between payoff and the propensity to imitate (or be imitated). For example, agents might copy the first person they see with a probability that depends negatively on their own payoffs and depends positively on the payoffs of those they seek to imitate.[3] In contrast to natural selection, the payoffs in such a model describe how people make choices, not on how rapidly they multiply, which is more consistent with the adaptive learning situations that we wish to study. However, individual payoffs must be observable by others for the model to make sense, an assumption that is not always satisfied.

3. *Reinforcement*. People tend to adopt actions that yielded a high payoff in the past, and to avoid actions that yielded a low payoff. This is the standard learning model in behavioral psychology, and it has increasingly captured the attention of economists.[4] As in imitative models, payoffs describe choice behavior, but it is one's *own* past payoffs that matter, not the payoffs of others. The basic premise is that the probability of taking an action in the present increases with the payoff that resulted from taking that action in the past. Models of this genre have been statistically estimated from the behavior of subjects playing simple games in laboratory settings, but the evidence is still too limited to draw general conclusions about their empirical validity.

4. *Best reply*. People adopt actions that optimize their expected payoff given what they expect others to do. This approach encompasses a variety of learning rules that attribute different degrees of "rationality" or "sophistication" to people's ability to forecast others' behavior. In the simplest such models, subjects choose best replies to the empirical frequency distribution of their opponents' previous actions. This is known as "fictitious play," a dynamic that we shall discuss more fully in the next section. Other versions posit more sophisticated rules by which agents update their beliefs about others' behavior.[5]

These four classes of learning rules are by no means exhaustive. One could postulate, for example, that people learn the same way neural nets do (Rumelhart and McClelland, 1986) or via genetic algorithms (Holland, 1975). We are not going to take a position about which of these most accurately represents adaptive behavior. If we had to take a position, we would guess that people adapt their adaptive behavior according to how they classify a situation (competitive or cooperative, for example) and reason from personal experience as well as from knowledge of others' experiences in analogous situations. In other words, learning can be very complicated indeed.

Nevertheless, considerable insight can be gained from studying the dynamics that arise from simple adaptive rules. For the sake of expositional clarity, we shall concentrate on best-reply dynamics. This choice is motivated by several considerations. First, best-reply dynamics (together with replicator dynamics) have been extensively studied and are well understood from a technical standpoint. Second, best-reply rules are more plausible than the other dynamics for the applications that we have in mind. Models of natural selection, for example, are premised on the idea that rules of behavior are genetically programmed and poorly adapted rules die out over the long run. This may well apply to certain types of metabehaviors—say, reasoning ability—but it is less compelling as a selection mechanism for particular behaviors. It is doubtful, for example, that going through a door first rather than second increases one's biological fitness, or that the strategy of going through doors is inherited. It is equally hard to believe that learning to type on different typewriter keyboards has significant implications for survival.

In contrast to the replicator model, reinforcement learning models specify how players make choices, and in this respect they are more suitable for modeling social and economic learning. Nevertheless, they are not entirely compelling. One drawback is that they take a very limited view of human rationality: agents are assumed to respond only to their own payoffs; they do not try to anticipate what others are going to do. A second problem has to do with applications: in real life, people may not play a given game often enough to learn from past payoffs. For example, it is doubtful that secretaries (or employers) make choices based on their past experience of choosing QWERTY or DVORAK; indeed, they face such a decision only once or twice in a lifetime.

A similar critique applies to models of imitation: while there is doubtless an imitative component to people's learning behavior, it does not seem credible that it is the *major* component. In adapting to interactive situations, it seems likely that people base their actions to a considerable extent on what they expect others to do, even though these expectations may be formed in a fairly primitive way. This is the foundational assumption of best-reply models. A particularly attractive feature of this framework is that it allows us to disentangle individuals' preferences from their beliefs about the world. In the models we shall consider, beliefs evolve dynamically as people interact and learn about the interactions of others; their preferences (what they choose given their beliefs) are fixed. This allows us to analyze a variety of issues that cannot be treated as coherently in the other frameworks, including the impact of having more or less information and the effects of risk aversion.

In principle, of course, it would be desirable to have a model of learning that incorporates elements of reinforcement, imitation, and best reply. To do so, however, would require a multidimensional representation of "payoffs," since payoffs mean different things in these models. This task is well beyond the scope of the present work. Our strategy is to show how a simple but plausible form of individual learning leads to a form of social learning, that is, to the evolution of predictable patterns of behavior at the societal level. Once the general architecture is in place, it will become clear how to apply it to a variety of other learning processes, including imitation and reinforcement.

2.2 Recurrent Games

In this section we define the basic situation that individuals are supposed to be learning about. An *n-person game* consists of n players $i = 1, 2, \ldots, n$, where each player i has a pure strategy space X_i and a utility function u_i that maps each n-tuple of strategies $x = (x_1, x_2, \ldots, x_n)$ to a payoff $u_i(x)$. In what follows, we shall assume that the sets X_i are finite, and we shall usually interpret the objects in X_i as actions that are publicly observable by others. (The term "action" is actually more restrictive than the term "strategy," which means a plan of action that may or may not be revealed in the course of the game. In most of what follows, however, pure strategies will correspond to observable actions.)

In standard game theory, each player is identified with a fixed individual, and if the game G is repeated for several periods, the same individuals always play it. This is known as a *repeated game*. In this book we shall be interested in games that are played repeatedly, but not necessarily by a fixed group of people. Instead, we shall think of the game as having *n roles*, and for each role $i = 1, 2, \ldots, n$ a nonempty *class* C_i of individuals who are eligible to play that role. Usually we shall assume that these classes are disjoint, though for symmetric games we shall sometimes assume that the players are drawn from a single class. In each discrete time period, $t = 1, 2, \ldots,$ one player is drawn at random from each of the n populations to play G. For the moment we can assume that all players are equally likely to be drawn; later we shall relax this hypothesis substantially. The elements $(X_i, u_i, C_i)_{1 \leq i \leq n}$ constitute a *recurrent game*.[6] The object of our study is the distribution of actions that people take in the n populations, and the way in which this distribution evolves over time.

2.3 Fictitious Play

We now introduce a model of how agents make choices. The essential idea is that each person constructs a simple statistical model of what other people are going to do, based on (possibly fragmentary) information about what they did in the past. The prototype of such learning models is fictitious play, which was originally proposed not as a learning rule but as a heuristic algorithm for computing Nash equilibria in certain classes of games (Brown, 1951; Robinson, 1951). The idea is quite natural: each player observes the actions that the others have chosen up to a given time t. Each player assumes that every other player is choosing according to a fixed probability distribution and that these distributions are independent among players. Thus the player in role i ("player i" for short) computes the observed frequency distribution of the actions taken by each player j up to time t, adopts this as a maximum likelihood estimate of the distribution player j is actually using, and then chooses a best reply to the product of these estimated distributions.

Specifically, let G be an n-person game with finite strategy space $X = \prod X_i$ and utility functions $\{u_i: i = 1, 2, \ldots, n\}$. The process unfolds in discrete time periods $t = 1, 2, 3, \ldots$ as follows. In period t, each "player" chooses an action, and the resulting n-tuple of actions is denoted by $x^t = (x_1^t, x_2^t, \ldots, x_n^t) \in X$. (For simplicity of exposition, we shall refer to the players as if they were fixed individuals, even though they are not.) The initial state is an arbitrary choice $x^1 \in X$. The history of play up through time t is $\bar{h}^t = (x^1, x^2, \ldots, x^t)$. For each i, let $p_i^t(x_i)$ be the proportion of the time that strategy x_i was played in the history $\bar{h}^t$. Thus for each i and each t we have $\sum_{X_i} p_i^t(x_i) = 1$. Let $p^t = \prod p_i^t$ be the associated product distribution. A *best reply* by player i given p^t is any strategy $x_i \in X_i$ that maximizes i's expected payoff, assuming that each player $j \neq i$ chooses according to the distribution p_j^t, where these distributions are independent. The *best-reply correspondence* is the set $BR_i(p^t)$ of i's best replies given p^t.

In general, let Δ_i denote the set of all probability distributions on the set X_i, and let $\Delta = \prod \Delta_i$ be the product set of distributions. For each $p \in \Delta$ and $x \in X$, the probability of x is $p(x) = \prod_i p_i(x_i)$. Let $X_{-i} = \prod_{j \neq i} X_j$ denote the space of actions taken by players other than i, and let x_{-i} denote an element of X_{-i}. Similarly, let $\Delta_{-i} = \prod_{j \neq i} \Delta_j$ be the product space of probability distributions for players $j \neq i$, and let p_{-i} denote an element of Δ_{-i}. Thus for each $p_{-i} \in \Delta_{-i}$ and each $x_{-i} \in X_{-i}$, $p_{-i}(x_{-i}) = \prod_{j \neq i} p_j(x_j)$. (Given an n-tuple $x \in X$, we shall also let x_{-i} denote x restricted to the coordinates $j \neq i$; similarly, given $p \in \Delta$, we let p_{-i} denote p restricted to the coordinates $j \neq i$.)

The utility of a probability distribution $p \in \Delta$ is assumed to be the expected utility of its outcomes. Hence it is convenient to extend u_i by defining $u_i(p) = \sum_{x \in X} u_i(x) p(x)$. We shall identify x_i with the distribution that places probability 1 on action x_i. For example, (x_i, p_{-i}) denotes a product distribution on X that places unit mass on x_i in the ith component. With these notational conventions, we can express the best-reply correspondence as follows:

$$BR_i(p^t) = \{x_i \in X_i\text{: for all } x_i' \in X_i, u_i(x_i, p_{-i}^t) \geq u_i(x_i', p_{-i}^t)\}. \tag{2.1}$$

A *fictitious play process* is a function $x^{t+1} = f(\bar{h}^t)$ that maps the history at each time t to an action-tuple x^{t+1} at time $t+1$, such that after some time t_0, each action is a best reply given the empirical distribution $p^t = p^t(\cdot \mid \bar{h}^t) \in \Delta$, that is,

$$\text{for all } t \geq t_0, x^{t+1} = f(\bar{h}^t) \text{ implies } x_i^{t+1} \in BR_i(p^t). \tag{2.2}$$

A *Nash equilibrium* of G is a product distribution $p^* \in \Delta$ such that

$$\text{for every } i, \text{ and every } p_i \in \Delta_i, u_i(p_i^*, p_{-i}^*) \geq u_i(p_i, p_{-i}^*). \tag{2.3}$$

A *pure Nash equilibrium* is a Nash equilibrium p^* such that $p^*(x) = 1$ for some particular $x \in X$. A Nash equilibrium p^* is *strict* if equation (2.3) holds strictly for every i and every $p_i \neq p_i^*$. Every strict Nash equilibrium is pure, because in a mixed Nash equilibrium some player has a continuum of best replies (given the others' strategies); hence (2.3) does not hold strictly.

A game G has the *fictitious play property* if every limit point of every sequence $\{p^t\}$ generated by a fictitious play process is a Nash equilibrium of G. In other words, every infinite convergent subsequence of a fictitious play process converges to a Nash equilibrium of G. This is weaker than requiring that every sequence $\{p^t\}$ generated by a fictitious play process converge to a Nash equilibrium. It is equivalent to saying that every such sequence converges to the closed *set* of Nash equilibria.

Various important classes of games have the fictitious play property. One family of two-person games with the property consists of finite zero-sum games. (A *zero-sum game* is a game in which the sum of payoffs to the players is zero for every choice of actions $x \in X$.)

THEOREM 2.1 (Robinson, 1951). *Every finite, zero-sum, two-person game has the fictitious play property.*

A second family of games with the fictitious play property comprises two-person games in which each player has just two actions and the payoffs satisfy the nondegeneracy condition described below. Let G be a two-person game with two actions for each player. (This is known as a 2×2 game.) Suppose that the payoffs from each pair of actions are as follows:

		Player 2	
		Action 1	Action 2
Player 1	Action 1	a_{11}, b_{11}	a_{12}, b_{12}
	Action 2	a_{21}, b_{21}	a_{22}, b_{22}

As usual, a_{ij} denotes the payoff to the row player (player 1) from the pair of actions (i, j), and b_{ij} is the payoff to the column player (player 2). The game G is *nondegenerate* if

$$a_{11} - a_{12} - a_{21} + a_{22} \neq 0, \quad b_{11} - b_{12} - b_{21} + b_{22} \neq 0. \tag{2.4}$$

This condition can be understood as follows. Suppose that G is nondegenerate and that G has a completely mixed equilibrium, where $(p, 1-p)$ are the probabilities that the row player puts on actions 1 and 2, and $(q, 1-q)$ are the probabilities that the column player puts on actions 1 and 2, $0 < p, q < 1$. The equilibrium conditions are

$$\begin{aligned} a_{11}q + a_{12}(1-q) &= a_{21}q + a_{22}(1-q), \\ b_{11}p + b_{21}(1-p) &= b_{12}p + b_{22}(1-p), \end{aligned}$$

which implies that

$$q = (a_{22} - a_{12})/(a_{11} - a_{12} - a_{21} + a_{22}) \text{ and } p = (b_{22} - b_{21})/(b_{11} - b_{12} - b_{21} + b_{22}).$$

Thus a nondegenerate game with a completely mixed equilibrium has a unique completely mixed equilibrium.

THEOREM 2.2 (Miyasawa, 1961; Monderer and Shapley, 1996a). *Every nondegenerate* 2×2 *matrix game has the fictitious play property.*[7]

To understand the dynamics of fictitious play, let us study its behavior in the etiquette game. For ease of exposition we shall relabel the actions A and B so that (A, A) and (B, B) are the coordination equilibria, and the payoff matrix has the form

	A	B
A	$1, \sqrt{2}$	$0, 0$
B	$0, 0$	$\sqrt{2}, 1$

At each time t, we shall represent the *state* by the pair $p^t = [(p_1^t, 1 - p_1^t), (p_2^t, 1 - p_2^t)] \in \Delta$, where $p_1^t \in [0, 1]$ is the proportion of the time that the row player played action A and $p_2^t \in [0, 1]$ is the proportion of the time that the column player played action A up through time t. Thus the numerical frequencies of A are tp_1^t and tp_2^t, respectively. For the sake of definiteness, we shall suppose that when A and B are both best replies for some player, the tie is broken in favor of A. For each state vector, $p = [(p_1, 1 - p_1), (p_2, 1 - p_2)]$ define the function $B^*(p) = [B_1^*(p), B_2^*(p)]$ as follows:

$$B_i^*(p) = (1, 0) \text{ iff action A is a best reply by } i; \quad \text{otherwise } B_i^*(p) = (0, 1). \tag{2.5}$$

Thus $B_i^*(p)$ is a vector-valued best-reply function with a particular tie-breaking rule, whereas $BR_i(p)$ is a *set* of pure-strategy best replies.

Let t_0 be the time after which fictitious play begins. Then for all $t \geq t_0$,

$$p^{t+1} = \frac{tp^t + B^*(p^t)}{t+1}. \tag{2.6}$$

The incremental motion is

$$\Delta p^{t+1} = p^{t+1} - p^t = \frac{B^*(p^t) - p^t}{t+1}. \tag{2.7}$$

When t is large, these incremental motions are small, and Δp^{t+1} is similar to a directional derivative. The instantaneous direction of motion of the corresponding continuous-time process $p(t)$ is:

$$\dot{p}(t) = dp(t)/dt = (1/t)[B^*(p(t)) - p(t)]. \tag{2.8}$$

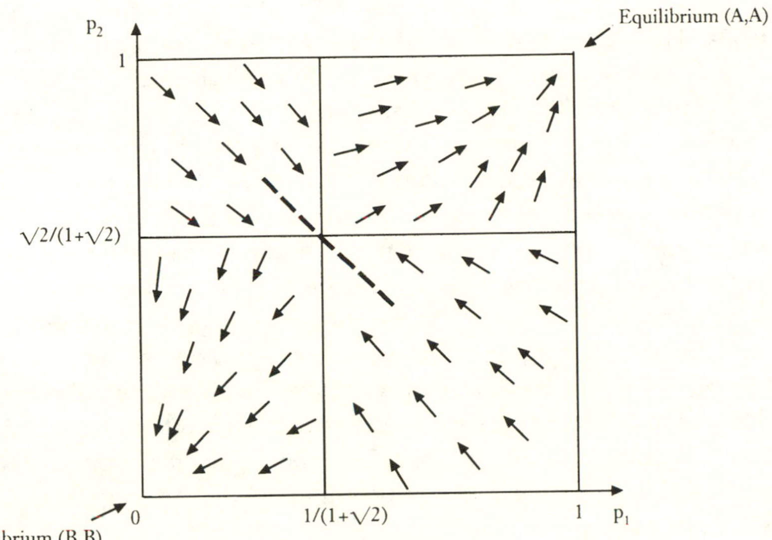

Figure 2.1. Vector field of the etiquette game.

Ignoring the scale factor $1/t$, which affects the speed but not the asymptotic behavior of the dynamics, we obtain the *continuous best-reply dynamic* (Matsui, 1992)[8]:

$$\dot{p}(t) = B^*(p(t)) - p(t). \tag{2.9}$$

Let us examine the behavior of the discrete-time process in greater detail. (The continuous-time version is similar.) The direction of motion at each point (the vector field) is illustrated for the etiquette game in Figure 2.1. From any point in the northeast rectangle, the process converges to (A, A), whereas from any point in the southwest rectangle, it converges to (B, B). From points in the northwest and southeast rectangles, the process moves toward the interior. Indeed, equation (2.7) shows that the motion at each such point is directed toward the opposite vertex of the square. Thus one of two things can happen: the process may eventually cross into the northeast or southwest rectangle, and from there converge to one of the pure equilibria; or (if it lies on the dashed line) the process moves back and forth between the northwest and southeast rectangles, gradually getting closer to the mixed Nash equilibrium but never landing on it.

It is important to observe, however, that while the frequency distribution p^t may be converging to the mixed equilibrium, the actual sequence of actions does not necessarily correspond to equilibrium play (Young, 1993a; Fudenberg and Kreps, 1993). Given the state p^t, the column player plays action A at time $t+1$ if and only if $\sqrt{2}p_1^t \geq 1 - p_1^t$. The row player chooses action A if and only if $\sqrt{2}(1 - p_2^t) \leq p_2^t$. These inequalities are in fact satisfied strictly because the p_i^t are always rational. (This is the reason for the payoff $\sqrt{2}$.) Suppose that the process begins in period 1 with a miscoordination, say, (A, B). Then $p_1^1 = 1 - p_2^1$ at the end of the first period. Thus they miscoordinate in the second period, which implies $p_1^2 = 1 - p_2^2$. Continuing in this manner, we see that in every period t, $p_1^t = 1 - p_2^t$; hence they miscoordinate again in period $t+1$. In terms of figure 2.1, the process is moving along the dashed line $p_1 + p_2 = 1$, sometimes hopping over the mixed equilibrium, and then eventually hopping back (because of the discrete-time jumps). But while the cumulative frequencies p^t are converging to the mixed equilibrium, the players are in fact miscoordinating in every period, and their payoffs are zero.

If the two players were actually playing the mixed equilibrium in each period, the row player would choose A and B with probabilities $1/(1+\sqrt{2})$ and $\sqrt{2}/(1+\sqrt{2})$, while the column player would independently choose A and B with probabilities $\sqrt{2}/(1+\sqrt{2})$ and $1/(1+\sqrt{2})$. Thus the probability of coordination would be $2\sqrt{2}/(1+\sqrt{2})^2$ and the expected payoff to each player would be $\sqrt{2}/(1+\sqrt{2})$. Thus, while the time-average behavior of the players in fictitious play is converging to the mixed equilibrium, this does not reflect what they are doing period-by-period.

2.4 Potential Games

Let us introduce another class of games for which fictitious play is well behaved, and that will play a role in future chapters.[9] Let G be an n-person game with finite strategy sets $X_1, X_2, \ldots, X_n$ and utility functions $u_i \colon X \to R$, where $X = \prod X_i$. G is a *(weighted) potential game* if there exists a function $\rho \colon X \to R$ and positive real numbers $\lambda_1, \lambda_2, \ldots, \lambda_n$ such that for every i, every $x_{-i} \in X_{-i}$, and every pair $x_i, x_i' \in X_i$,

$$\lambda_i u_i(x_i, x_{-i}) - \lambda_i u_i(x_i', x_{-i}) = \rho(x_i, x_{-i}) - \rho(x_i', x_{-i}). \tag{2.10}$$

In other words, the utility functions can be rescaled so that whenever a player changes strategy unilaterally, the change in his or her utility equals the change in potential. The point here is that the same potential function ρ applies to all players. In particular, suppose that the players choose some n-tuple of actions $x^0 \in X$. A *path of unilateral deviations* (UD-*path*) is a sequence of form $x^0, x^1, \ldots, x^k$, where each $x^j \in X$, and x^j differs from its predecessor x^{j-1} in exactly one component. Given distinct n-tuples x and x', there will typically exist multiple UD-paths from x to x'. If G is a potential game, the total change in (weighted) utility along such a path, summed over all deviating players, must be the same independently of the path taken.

The most transparent example of a potential game is a game with identical payoff functions $u_i(x) = u(x)$ for every player i. A considerably less obvious example is the following class of Cournot oligopoly games. Consider n firms with identical cost functions $c(q_i)$, where $c(q_i)$ is continuously differentiable in q_i and q_i is the quantity produced by firm i. Let $Q = \sum q_i$, and suppose the inverse demand function takes the form $P = a - bQ$, where P is the unit price and $a, b > 0$. Then the profit for firm i (which we identify with i's payoff) is

$$u_i(q_1, q_2, \ldots, q_n) = (a - bQ)q_i - c(q_i).$$

It is straightforward to check that the following is a potential function for this game:

$$\rho(q) = a\sum_{i=1}^{n} q_i - b\sum_{i=1}^{n} q_i^2 - b\sum_{1\leq i<j\leq m} q_i q_j - \sum_{i=1}^{n} c_i(q_i).^{10} \tag{2.11}$$

As a third example, we claim that every symmetric 2×2 game is a potential game. Indeed, the payoffs from such a game can be written in the form

	A	B
A	a, a	c, d
B	d, c	b, b

. (2.12)

It is easy to see that a potential function for this game is

$$\begin{aligned} &\rho(\mathrm{A}, \mathrm{A}) = a - d \qquad &&\rho(\mathrm{A}, \mathrm{B}) = 0 \\ &\rho(\mathrm{B}, \mathrm{A}) = 0 \qquad &&\rho(\mathrm{B}, \mathrm{B}) = b - c. \end{aligned} \tag{2.13}$$

We claim that every weighted potential game possesses at least one Nash equilibrium in pure strategies. To see this, consider any strategy-tuple $x^0 \in X$. If x^0 is not a Nash equilibrium of G, there exists a player i who can obtain a higher payoff by changing to some strategy $x_i^1 \neq x_i^0$. Continuing in this manner, one generates a *finite improvement path*, namely, a UD-path $x^0, x^1, \ldots, x^k$ such that the unique deviator at each stage strictly increases his utility. It follows that the potential is strictly increasing along any finite improvement path. Since the strategy space X is finite, the path must end at some $x^* \in X$ from which no further deviations are worthwhile; that is, x^* is a pure Nash equilibrium of G.[11] The following result shows that in a potential game, fictitious play always converges to the set of Nash equilibria (whether pure or mixed).

THEOREM 2.4 (Monderer and Shapley, 1996b). *Every weighted potential game has the fictitious play property.*

2.5 Nonconvergence of Fictitious Play

Many games do not have the fictitious play property, however. The first such example is due to Shapley (1964), and can be motivated as follows.

The Fashion Game

Individuals 1 and 2 go to the same parties, and each pays attention to what the other wears. Player 1 is a fashion follower and likes to wear the same color that player 2 is wearing. Player 2 is a fashion leader and likes to wear something that contrasts appropriately with what player 1 is wearing. For example, if player 1 is wearing red, player 2 prefers to wear blue (but not yellow); if player 1 is wearing yellow, player 2 prefers red, while if player 1 dons blue, player 2 wants to be in yellow. The payoff matrix has the following form:

		Player 2		
		Red	Yellow	Blue
	Red	1, 0	0, 0	0, 1
Player 1	Yellow	0, 1	1, 0	0, 0
	Blue	0, 0	0, 1	1, 0

This game has a unique Nash equilibrium in which the players take each action with probability 1/3.

Assume now that both players use fictitious play as a learning rule, and that they both start by wearing red in the first period. The process then unfolds as follows (we assume that ties are broken according to the hierarchy yellow > blue > red).

	t													
	1	2	3	4	5	6	7	8	9	10	11	12	13	14 . . .
Player 1	R	R	B	B	B	B	B	Y	Y	Y	Y	Y	Y	Y
Player 2	R	B	B	B	Y	Y	Y	Y	Y	Y	Y	Y	Y	R

Roughly speaking, the process follows the fashion cycle red → blue → yellow → red, with player 2 leading the cycle, and player 1 following. More precisely, define a *run* to be a sequence of consecutive periods in which neither player changes action. The runs in the above sequence follow the cyclic pattern shown below.

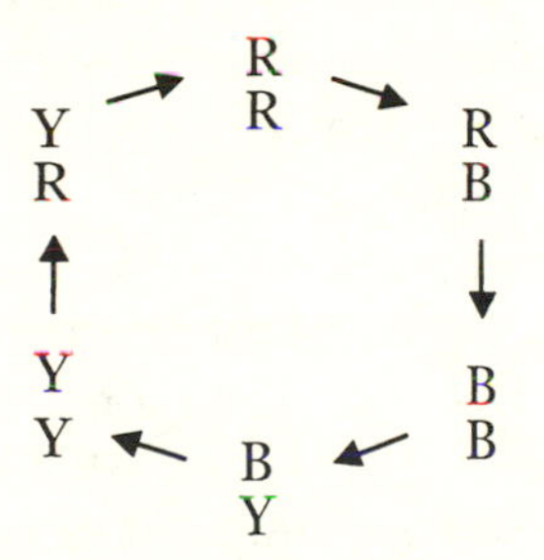

Moreover, the number of periods in each consecutive cycle grows exponentially. Let $p_i^t = (p_{iR}^t, p_{iY}^t, p_{iB}^t)$ be the relative frequency with which player i ($i = 1, 2$) plays each strategy up through time t. Shapley (1964) showed that (p_1^t, p_2^t) does not converge to the Nash equilibrium; instead, it converges to a cycle in the product space $\Delta_3 \times \Delta_3$, where $\Delta_3 = \{(q_1, q_2, q_3) \in R_+^3 : \sum q_i = 1\}$.

While this example might suggest that fictitious play has its drawbacks as a learning rule, it also suggests that equilibrium has its drawbacks as a solution concept. Fashion would not be fashion if it were not constantly in flux. It is a form of interaction in which we *expect* cyclic behavior.[12] The fact that fictitious play cycles in such a game might be seen as an argument in its favor, since cycling in this case is at least as plausible as Nash equilibrium behavior.

There are other games, however, in which Nash equilibrium is more clearly the "right" solution, yet fictitious play fails to find it. Define a two-person *coordination game* to be a game in which each player has the same number of actions, which can be indexed so that it is a strict Nash equilibrium for the players to play actions having the same index. In other words, let the actions of the players be indexed $j = 1, 2, \ldots, m$, and let the payoff from playing the pair of actions with indices (j, k) be a_{jk} for the row player and b_{jk} for the column player. In a coordination game, $a_{jj} > a_{kj}$ and $b_{jj} > b_{jk}$ for every distinct pair of indices j and k. Consider the following example, drawn from Foster and Young (1998).

The Merry-Go-Round Game

Rick and Cathy are in love, but they are not allowed to communicate. Once a day, at an appointed time, they can take a ride on a merry-go-round, which has nine pairs of horses (Figure 2.2). Before taking a ride, each chooses one of the pairs without communicating his or her choice to the other. There are no other riders. If they book the same pair they get to ride side by side, which is their preferred outcome. If they choose different pairs, their payoffs depend on how conveniently they can look at each other. If Rick sits just behind Cathy, then he can see her but she has trouble looking at him, because the horses all face in the clockwise direction. Say this outcome has payoff 4 for Rick and payoff 0 for Cathy. If they are on opposite sides of the circle, they can both see each other easily, but the one who has to crane his or her neck less has a slightly higher payoff (5 compared to 4). If they sit side by side, they can look at each other to their hearts' content, which has a payoff of 6 for both. The payoff matrix is therefore as follows:

6, 6	4, 0	4, 0	4, 0	5, 4	4, 5	0, 4	0, 4	0, 4
0, 4	6, 6	4, 0	4, 0	4, 0	5, 4	4, 5	0, 4	0, 4
0, 4	0, 4	6, 6	4, 0	4, 0	4, 0	5, 4	4, 5	0, 4
0, 4	0, 4	0, 4	6, 6	4, 0	4, 0	4, 0	5, 4	4, 5
4, 5	0, 4	0, 4	0, 4	6, 6	4, 0	4, 0	4, 0	5, 4
5, 4	4, 5	0, 4	0, 4	0, 4	6, 6	4, 0	4, 0	4, 0
4, 0	5, 4	4, 5	0, 4	0, 4	0, 4	6, 6	4, 0	4, 0
4, 0	4, 0	5, 4	4, 5	0, 4	0, 4	0, 4	6, 6	4, 0
4, 0	4, 0	4, 0	5, 4	4, 5	0, 4	0, 4	0, 4	6, 6

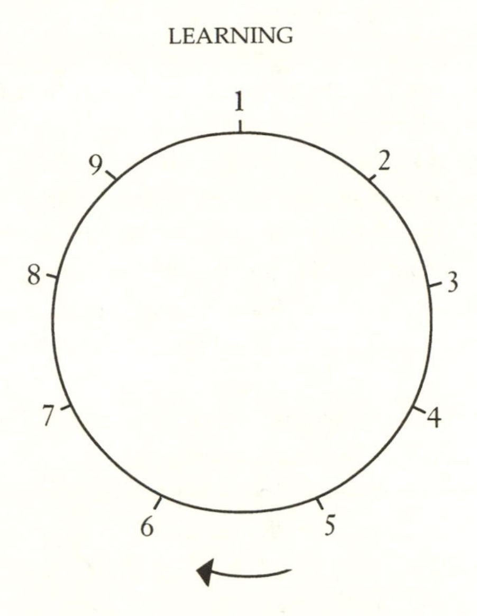

Figure 2.2. The merry-go-round game.

A similar story can be told for two players who want to coordinate on some date in the calendar. For example, think of two car manufacturers who announce their new models once a year. If they announce on the same date, they generate a lot of media coverage. If one announces "earlier" than the other does, they receive less publicity in total, and the earlier announcement casts a "publicity shadow" on the later one. In other words, miscoordination hurts both parties but it hurts one party more than the other.

When players try to learn such a game by fictitious play, they can fall into a cyclic pattern in which they keep locating on (more or less) opposite sides of the circle, and never coordinate. That is, under certain initial conditions, the distribution of the other person's previous actions are such that each player would rather advance one position (to get a better view of the other) than try to coordinate exactly, which carries the risk of "missing" by one position.

2.6 Adaptive Play

One might wonder why the players are not clever enough to figure out that they are caught in a cycle. Perhaps if they analyzed the data in a more sophisticated way, they would avoid falling into such a trap. We shall not pursue this avenue further here, although it is certainly one that should be pursued. Let us recall, however, that the situation we have in mind is a recurrent game played by a changing cast of characters who have limited information, have only modest reasoning ability, are short-sighted, and sometimes do unexplained or even foolish things. Such players may be quite unable to make sophisticated forecasts; nevertheless they may be able to feel their way collectively toward interesting and even sensible solutions. Indeed, this is the case when the players are even *less* rational and *less* well-informed than they are in fictitious play.

To make these ideas precise, let us fix an n-person game with finite strategy sets $X_1, \ldots, X_n$. The joint strategy space is $X = \prod X_i$, and the payoff functions are u_i: $X \to R$. In each period, one player is drawn at random from each of n disjoint *classes* $C_1, C_2, \ldots, C_n$, to play the game G. Let $x_i^t \in X_i$ denote the *action* taken by the i player at time t. The *play*, or *record*, at time t is the vector $x^t = (x_1^t, x_2^t, \ldots, x_n^t)$. The *state* of the system at the end of period t is the sequence of the last m plays: $h^t = (x^{t-m+1}, \ldots, x^t)$. The value of m determines how far back in time the players are willing (or able) to look.

Let X^m denote the set of all states, that is, the set of all histories of length m. The process starts out in an arbitrary state h^0 consisting of m records. Given that the process is in state h^t at the end of period t, consider some player who is selected to play role i in period $t+1$. Each player has access to information about what some of the previous players have done, which is relayed through established networks of friends, neighbors, co-workers, and so on. In other words, access to information is part of an agent's situation, not the result of an optimal search. We model the information transmission process as a random variable: the current i player draws a sample of size s from the set of actions taken by each of the other role players over the last m periods. These $n-1$ samples are drawn independently. Let $\hat{p}_{ij}^t \in \Delta_j$ denote the sample proportion of j's actions in i's sample, and let $\hat{p}_{-i}^t = \prod_{j \neq i} \hat{p}_{ij}^t$. (A "hat" over a variable indicates that it is a random variable.)

Idiosyncratic behaviors are modeled as follows. Let $\varepsilon > 0$ be a small positive probability called the *error rate*. With probability $1-\varepsilon$ the i player chooses a best reply to $\hat{p}_{-i}^t$ and with probability ε he chooses a strategy in X_i at random. If there are ties in best reply, we assume each is chosen with equal probability.[13] These errors correspond to mutations in biological models of evolution. Here, they are interpreted as minor disturbances to the adaptive process, which include exogenous stochastic shocks as well as unexplained variation (idiosyncracies) in personal behavior. The probability of making an error is assumed to be independent across players. Taken together, these factors define a Markov process called *adaptive play* with memory m, sample size s, and error rate ε (Young, 1993a). This model has four key features: *boundedly rational responses* to forecasts of others' recent behavior, which are estimated from *limited data* and are perturbed by *stochastic shocks*. In the following chapters, we shall examine the properties of this process in detail, and show how it leads to a new concept of equilibrium (and disequilibrium) selection in games.

Chapter 3

DYNAMIC AND STOCHASTIC STABILITY

3.1 Asymptotic Stability

Before analyzing the dynamic behavior of the learning model introduced in the preceding chapter, we need a framework for thinking about the asymptotic properties of dynamic systems in general. Let Z denote a set (finite or infinite) of possible states of a dynamical system. For analytical convenience, we shall assume that Z is contained in a finite Euclidean space R^k. A *discrete-time dynamical process* is a function $\zeta(t, z)$ defined for all discrete times $t = 0, 1, 2, 3, \ldots$ and all z in Z such that if the process is in state z at time t, then it is in state $z' = \zeta(t, z)$ at time $t + 1$. From any initial point z^0 the *solution path* of the process is $\{z^0, z^1 = \zeta(0, z^0), z^2 = \zeta(1, z^1), \ldots\}$.

There are two senses in which a state z can be said to be "stable" in such a process. The weaker condition (Lyapunov stability) says that there is no local push away from the state; the stronger condition (asymptotic stability) says that there is a local pull toward the state. Formally, a state z is *Lyapunov stable* if every open neighborhood $\mathcal{B}$ of z contains an open neighborhood $\mathcal{B}^0$ of z such that any solution path that enters $\mathcal{B}^0$ stays in $\mathcal{B}$ from then on. In particular, the process need not return to z, but if it comes close to z it must remain close. A state z is *asymptotically stable* if it is Lyapunov stable and the open neighborhood $\mathcal{B}^0$ can be chosen so that if a solution path ever enters $\mathcal{B}^0$, then it converges to z.

Although most of the processes considered in this book take place in discrete time, we shall briefly outline the analogous ideas for continuous-time processes. Let Z be a compact subset of Euclidean space R^k. A *continuous-time dynamical process* is a continuous mapping $\xi\colon R \times Z \to Z$ where for all $z^0 \in Z$, $\xi(0, z^0) = z^0$ and $\xi(t, z^0)$ represents the position of the process at time t given that it started at z^0. Such a system often arises as the solution to a system of first-order differential equations of the form

$$\dot{z} = \phi(z) \text{ where } \phi\colon Z \to R^k. \tag{3.1}$$

Here ϕ defines the *vector field*, that is, the direction and velocity of motion at each point in Z. A *solution* to (3.1) is a function $\xi: R \times Z \to Z$ such that

$$d\xi(t,z)/dt = \phi[\xi(t,z)] \text{ for all } t \in R \text{ and } z \in Z. \tag{3.2}$$

The function ϕ is *Lipschitz continuous* on a domain Z^* if there exists a nonnegative real number L (the Lipschitz constant) such that $|\phi(z) - \phi(z')| \leq L|z - z'|$ for all $z, z' \in Z^*$. A fundamental theorem of dynamical systems states: *If ϕ is Lipschitz continuous on an open domain Z^* containing Z, then for every $z^0 \in Z$ there exists a unique solution ξ such that $\xi(0, z^0) = z^0$; moreover $\xi(t, z^0)$ is continuous in t and z^0.*[1]

To illustrate asymptotic stability in the discrete case, consider the dynamics defining fictitious play:

$$\Delta p^{t+1} = \frac{B^*(p^t) - p^t}{t+1}. \tag{3.3}$$

Although we derived this equation for 2×2 games, the same equation holds for any finite n-person game G with an appropriate rule for resolving best-reply ties.

Let $x^* \in X$ be a strict Nash equilibrium, that is, x_i^* is the unique best reply to x_{-i}^* for every player i. Since the utility functions u_i: $\Delta \to R$ are continuous, there exists an open neighborhood $\mathcal{B}_{x^*}$ of x^* such that x_i^* is the unique best reply to p_{-i} for all $p \in \mathcal{B}_{x^*}$. From this and (3.3), it follows that *every strict Nash equilibrium is asymptotically stable in the discrete-time fictitious play dynamic.* The same holds for the continuous-time version of the best-reply dynamic (2.9). In particular, every coordination equilibrium of a coordination game is asymptotically stable in the discrete (or continuous) form of the best-reply dynamic.

A similar result holds for a much larger class of learning dynamics, including the replicator dynamic. Let $\Delta = \prod \Delta_i$ be the space of mixed strategies for a finite n-person game G. Imagine that each population of players forms a continuum, and that at each time t, each player has an action, which he or she plays against randomly drawn members of the other populations. We can represent the *state* at time t by a vector $p(t) \in \Delta$, where $p_{ik}(t)$ denotes the proportion of population i playing the "kth" action in X_i and k indexes the actions in X_i. A *growth rate dynamic* is a dynamical system of form

$$\dot{p}_{ik} = g_{ik}(p)p_{ik} \text{ where } p_i \cdot g_i(p) = 0 \text{ and } g_i\text{: } \Delta \to \Delta_i. \tag{3.4}$$

The condition $p_i \cdot g_i(p) = 0$ is needed to keep the process in the simplex Δ_i for each i. Let us assume that, for every i, g_i is Lipschitz continuous on an open domain containing Δ, so that the system has a unique solution from any initial state p^0. (Such a process is sometimes called a *regular selection dynamic*.) The function $g_{ik}(p)$ expresses the rate at which the proportion of population i playing the kth action in X_i increases (or decreases) as a function of the current state $p \in \Delta$. Note that an action cannot grow if it is not present in the population. A particular consequence is that every vertex of Δ (every n-tuple $x \in X$) is a rest point of the dynamic.

Let x_{ik} denote the kth action of player i. The dynamic is *payoff monotonic* if in each state, the growth rates of the various actions are strictly monotonic in their expected payoffs.[2] In other words, for all $p \in \Delta$, all i, and all $k \neq k'$,

$$u_i(x_{ik}, p_{-i}) > u_i(x_{ik'}, p_{-i}) \text{ iff } g_{ik}(p) > g_{ik'}(p). \tag{3.5}$$

The *replicator dynamic* is the special case in which the growth rate of each action x_{ik} equals the amount by which its payoff exceeds the average payoff to members of population i, that is,

$$\dot{p}_{ik} = [u_i(x_{ik}, p_{-i}) - u_i(p)]p_{ik}. \tag{3.6}$$

It can be shown that *in a payoff monotonic, regular selection dynamic, every strict Nash equilibrium is asymptotically stable*. Thus if the game has multiple strict Nash equilibria (as in a coordination game), the dynamics can "select" any one of them, depending on the initial conditions.[3]

3.2 Stochastic Stability

Even though asymptotic stability is essentially a deterministic concept, it is sometimes invoked as a criterion of stability in the presence of random perturbations. To illustrate, suppose that a process is in an asymptotically stable state and that it receives a small, one-time shock. If the shock is sufficiently small, then by the definition of asymptotic stability the process will eventually revert to its original state. Since the impact of the shock ultimately wears off, one could say that the state is stable in the presence of stochastic perturbations. The difficulty with this argument is that, in reality, stochastic perturbations are not isolated events. What happens when the system receives a second shock before it has recovered from the first, then receives a third shock before it has recovered from the second, and so forth? The concept of asymptotic stability is inadequate to answer this question.

In the next two sections, we introduce an alternative concept of stability that takes into account the persistent nature of stochastic perturbations. Instead of representing the process as a deterministic system that is occasionally tweaked by a random shock, we shall represent it as a stochastic dynamical system, in which perturbations are incorporated directly into the equations of motion. Such perturbations are a natural feature of most adaptive processes, including those discussed in the preceding chapter. For example, the replicator dynamic amounts to a continuous approximation of a finite process in which individuals from a finite population reproduce (and die) with *probabilities* that depend on their payoffs from interactions. In models of imitation, the random component follows from the assumption that individuals update their strategies with probabilities that are payoff dependent. In reinforcement models, agents are assumed to choose actions with probabilities that depend on the payoffs they received from these actions in the past. And so forth. Virtually any plausible model of adaptive behavior we can think of has stochastic components.

Although previous authors have recognized the importance of stochastic perturbations in evolutionary models, it has been standard to suppress them on the grounds that they are inherently small, or else small in the aggregate because the variability is averaged over many individuals, and to study the expected motion of the process *as if* it were a deterministic system. While this may be a reasonable approximation of the short-run (or even medium-run) behavior of the process, however, it may be a very poor indicator of the long-run behavior of the process. A remarkable feature of stochastic dynamical systems is that their long-run (asymptotic) behavior can differ radically from the corresponding deterministic process *no matter how small the noise term is*. The presence of noise, however minute, can qualitatively alter the behavior of the dynamics. But there is also an unexpected dividend: since such processes are often ergodic, their *long-run average* behavior can be predicted much more sharply than that of the corresponding deterministic dynamics, whose motion usually depends on the initial state. In the next several sections we shall develop a general framework for analyzing these issues. We shall then apply it to adaptive learning, and show how it leads to a theory of equilibrium (and disequilibrium) selection for games in general.

3.3 ELEMENTS OF MARKOV CHAIN THEORY

A *discrete-time Markov process* on a finite state space Z specifies the probability of transiting to each state in Z in the next period, given that the process is currently in some given state z. Specifically, for every pair of states $z, z' \in Z$, and every time $t \geq 0$, let $P_{zz'}(t)$ be the *transition probability* of moving to state z' at time $t+1$ conditional on being in state z at time t. If $P_{zz'} = P_{zz'}(t)$ is independent of t, we say that the transition probabilities are *time homogeneous*. This assumption will be maintained throughout the following discussion.

Suppose that the initial state is z^0. For each $t > 0$, let $\mu^t(z \mid z^0)$ be the *relative frequency* with which state z is visited during the first t periods. It can be shown that as t goes to infinity, $\mu^t(z \mid z^0)$ converges almost surely to a probability distribution $\mu^\infty(z \mid z^0)$, called the *asymptotic frequency distribution* of the process conditional on z^0.[4] The distribution μ^∞ can be interpreted as a selection criterion: over the long run, the process *selects* those states on which $\mu^\infty(z \mid z^0)$ puts positive probability. Of course, the state or states that are selected in this sense may depend on the initial state, z^0. If so, we say that the process is *path dependent*, or *nonergodic*. If the asymptotic distribution $\mu^\infty(z \mid z^0)$ is independent of z^0, the process is *ergodic*.

Consider a finite Markov process on Z defined by its transition matrix, P. A state z' is *accessible* from state z, written $z \to z'$, if there is a positive probability of moving from z to z' in a finite number of periods. States z and z' *communicate*, written $z \sim z'$, if each is accessible from the other. Clearly $\sim$ is an equivalence relation, so it partitions the space Z into equivalence classes, which are known as *communication classes*. A *recurrent class* of P is a communication class such that no state outside the class is accessible from any state inside it. It is straightforward to show that every finite Markov chain has at least one recurrent class. Denote the recurrent classes by $E_1, E_2, \ldots, E_K$. A state is *recurrent* if it is contained in one of the recurrent classes; otherwise it is *transient*. A state is *absorbing* if it constitutes a singleton recurrent class. If the process has exactly one recurrent class, which consists of the whole state space, the process is said to be *irreducible*. Equivalently, *a process is irreducible if and only if there is a positive probability of moving from any state to any other state in a finite number of periods.*

The asymptotic properties of finite Markov chains can be studied algebraically as follows. Let $z_1, \ldots, z_N$ be an enumeration of the states, and let μ be a probability distribution on Z written out as a row vector: $\mu = (\mu(z_1), \mu(z_2), \ldots, \mu(z_N))$. Consider the system of linear equations

$$\mu P = \mu, \text{ where } \mu \geq 0 \text{ and } \sum_{z \in Z} \mu(z) = 1. \tag{3.7}$$

It can be shown that this system always has at least one solution μ, which is called a *stationary distribution* of the process P. The term *stationary* is apt for the following reason. Suppose that $\mu(z)$ is the probability of being in state z at some time t. The probability of being in state z at time $t+1$ is the probability (summed over all states w) of being in state w at time t and transiting to z in the next period. In other words, the probability of being in state z in period $t+1$ equals $\sum_w \mu(w) P_{wz}$. Equation (3.7) states that this probability is precisely $\mu(z)$ for every $z \in Z$. In other words, if μ is the probability distribution over states at some time t, then μ is also the probability distribution over states at all subsequent times.

The stationarity equation (3.7) has a unique solution if and only if P has a unique recurrent class, in which case we refer to *the* stationary distribution μ of P. A fundamental result for finite Markov chains states: *If P has a unique recurrent class, then the stationary distribution μ describes the time-average asymptotic behavior of the process independently of the initial state z^0:*

$$\lim_{t \to \infty} \mu^t(z \mid z^0) = \mu^\infty(z \mid z^0) = \mu(z). \tag{3.8}$$

By contrast, if P has more than one recurrent class, then it is nonergodic, and the asymptotic distribution depends on where the process starts.

The learning processes we shall consider have a further property, which allows us to make even sharper statements about their asymptotic behavior. Let P be a finite Markov process on the set Z, and for each state z, let N_z be the set of all integers $n \geq 1$ such that there is a positive probability of moving from z to z in exactly n periods. The process is *aperiodic* if for every z, the greatest common divisor of N_z is unity.

For a process that is aperiodic and irreducible, not only is it true that the time-average behavior of the process converges to the unique stationary distribution μ; its position at each point in time t is also approximated by μ when t is sufficiently large. More precisely, let $\nu^t(z \mid z^0)$

be the probability that the process is in state z at time t, given that the process was in state z^0 at time $t = 0$. (By contrast, $\mu^t(z \mid z^0)$ is the frequency distribution over the first t periods, conditional on beginning in z^0.) Letting P^t be the t-fold product of P, it follows that

$$\nu^t(z \mid z^0) = P^t_{z^0 z}. \tag{3.9}$$

If the process is irreducible and aperiodic, it can be shown that P^t converges to the matrix P^∞ in which every row equals μ. Thus we have

$$\lim_{t\to\infty} \nu^t(z \mid z^0) = \mu(z) \text{ for all } z \in Z, \tag{3.10}$$

where μ is the unique stationary distribution. It follows that in every aperiodic, irreducible, finite Markov process, the probability $\nu^t(z \mid z^0)$ of being in state z *at a given time* t is essentially the same as the probability $\mu^t(z \mid z^0)$ of being in state z *up to time* t when t is large. That is, with probability one, both $\nu^t(z \mid z^0)$ and $\mu^t(z \mid z^0)$ converge to $\mu(z)$ independently of the initial state.

These ideas can be illustrated with the learning model introduced in the previous chapter. Let G be a finite n-person game with strategy space $X = \prod X_i$. Consider adaptive learning with memory m, sample size s, and error rate ε. This is clearly a finite Markov chain: the state space $Z = X^m$ consists of all length-m histories, and for each pair of histories $h, h' \in X^m$, there is a fixed probability $P_{hh'}$ of transiting from h to h' in one period. We shall denote this process by its transition matrix, $P^{m,s,\varepsilon}$.

We claim that when the error rate ε is positive, the process is irreducible. To establish this, suppose that the process is in state $h = (x^{t-m+1}, \ldots, x^t)$ at the end of period t. Because of random error, there is a positive probability of generating an arbitrary set of actions in the next period. Thus within m periods, there is a positive probability of reaching any other history of length m. It follows that the process is irreducible and its asymptotic distribution is the unique solution to (3.7). (Note also that the process is aperiodic.) We shall denote this distribution by $\mu^{m,s,\varepsilon}$, or sometimes by μ^ε, when m and s are understood.

When the error rate is zero, by contrast, the process may be *reducible*. To see this, suppose that G has a strict pure strategy Nash equilibrium x^*. Consider the history in which x^* was played m times in succession: $h^* = (x^*, x^*, \ldots, x^*)$. For any values of s and m such that $1 \le s \le m$, h^* is an absorbing state of the process, because for every i, x_i^* is the unique best reply to every set of samples that i might draw from the history h^*. Such a state h^* is called a *convention*. It is a situation in which, for as long as anyone can remember, people have always played x^*. Assuming that everyone else continues to adhere to the convention, it is in each person's interest to adhere to it also, and thus the convention stays in place—it is a self-enforcing mechanism of social coordination. The proof of the following statement is straightforward and is left to the reader.

The only absorbing states of the process $P^{m,s,0}$ are the conventions, that is, states of form $(x^*, x^*, \ldots, x^*)$ where x^* is a strict Nash equilibrium of G.

Any game with more than one strict Nash equilibrium has multiple absorbing states; hence the learning process $P^{m,s,0}$ is reducible and therefore not ergodic. To illustrate, consider the typewriter game with the following payoff matrix:

	D	Q
D	5, 5	0, 0
Q	0, 0	4, 4

Choose the parameters in adaptive learning as follows: memory, $m = 4$; sample size, $s = 2$; and error rate, $\varepsilon = 0$. Each state can be represented by a 2×4 block of Qs and Ds, where the top line represents the row player's previous four actions and the bottom line represents the column player's previous four actions. For example, the state

DQDQ
QDDQ

means that the row player chose D four periods ago, Q three periods ago, D two periods ago, and Q one period ago. Similarly, the column player chose Q four periods ago, D three and two periods ago, and Q one period ago. The probability of transiting to other states depends on the samples that the agents happen to choose from this history. (For convenience we assume that all samples are equally likely to be drawn.) Once the new actions are played, the leftmost (oldest) actions are deleted and the new actions are adjoined to the right. Since coordinating on D has a higher payoff than does coordinating on Q, D is the unique best reply to all samples of size two *except* the sample {Q, Q}. Therefore the

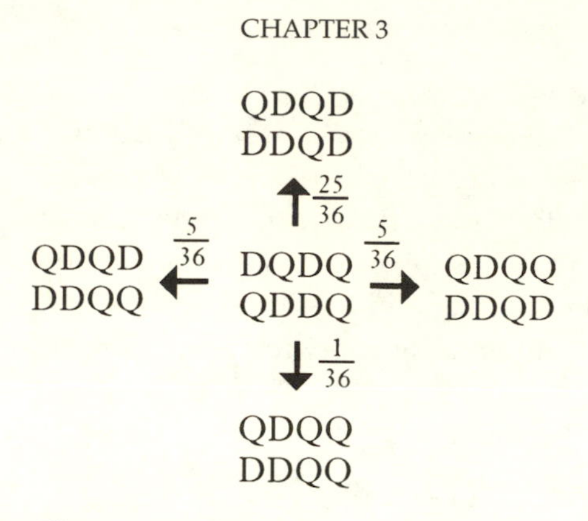

Figure 3.1. Transition probabilities to neighboring states.

one-period transition probabilities from the above state are those shown in Figure 3.1.

It is cumbersome to write down all 256 states, and even more cumbersome to solve the stationarity equations (3.7). Fortunately we can say a good deal about the asymptotic behavior of the process without taking this step. We already know that the process has two absorbing states: the conventions all-D and all-Q. The first question we need to address is whether these constitute the only recurrent classes. Note that this does not follow immediately, because there could exist a recurrent class that contains multiple states, none of which is absorbing. In the preceding example, however, this cannot happen because from any state there is a positive probability of moving in a finite number of periods to the all-D or all-Q state. (The verification is left to the reader.) It follows that all-D and all-Q are the only recurrent classes, and that all other states are transient. The process is nonergodic in the sense that the asymptotic distribution—the probabilities that all-D versus all-Q will eventually be reached—depends on the state in which the process starts. In other words, from any initial state, the process will end up in all-D or all-Q with probability one. However, the process is nonergodic because the probability that it will end up in all-D or all-Q depends on the starting position. Figure 3.2 shows two routes (among many) from a given starting state to all-D on the one hand and to all-Q on the other. In this sense, the long-run behavior of the process is only weakly predictable.

When the error rate is positive, by contrast, the process is ergodic and its long-run average behavior does *not* depend on the initial conditions. In particular, the stationary distribution gives the (approximate)

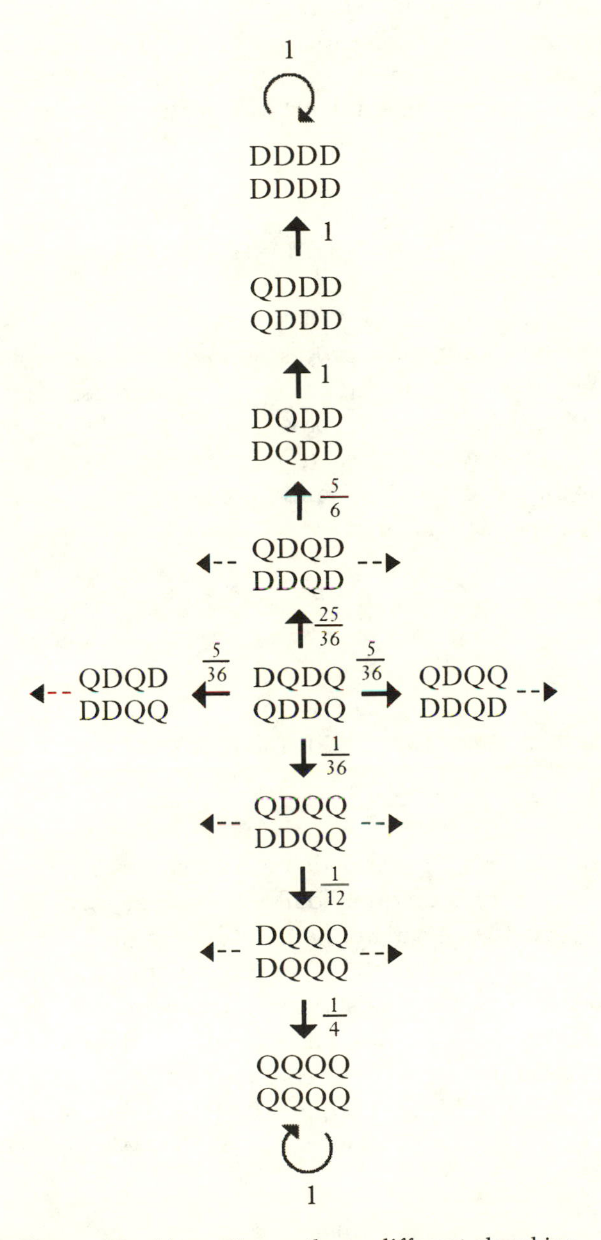

Figure 3.2. Alternative paths to different absorbing states.

likelihood that the process will be in each state z at each future date t independently of the initial state, as long as t is large. In most cases, solving for the stationary distribution by algebraic methods is impracticable because of the large number of states. In the next section we show how to get a handle on the stationary distribution by a different technique.

3.4 Perturbed Markov Processes

Consider a Markov process P^0 defined on a finite state space Z. A *perturbation* of P^0 is a Markov process whose transition probabilities are slightly perturbed or distorted versions of the transition probabilities $P^0_{zz'}$. Specifically, for each ε in some interval $[0, \varepsilon^*]$ let P^ε be a Markov process on Z. We say that P^ε is a *regular perturbed Markov process* if P^ε is irreducible for every $\varepsilon \in (0, \varepsilon^*]$, and for every $z, z' \in Z$, $P^\varepsilon_{zz'}$ approaches $P^0_{zz'}$ at an exponential rate, that is,

$$\lim_{\varepsilon \to 0} P^\varepsilon_{zz'} = P^0_{zz'}, \tag{3.11}$$

and

$$\text{if } P^\varepsilon_{zz'} > 0 \text{ for some } \varepsilon > 0, \text{ then } 0 < \lim_{\varepsilon \to 0} P^\varepsilon_{zz'}/\varepsilon^{r(z,z')} < \infty$$
$$\text{for some } r(z, z') \geq 0. \tag{3.12}$$

The real number $r(z, z')$ is called the *resistance* of the transition $z \to z'$. Note that $r(z, z')$ is uniquely defined, because there cannot be two distinct exponents that satisfy the condition in (3.12). Note also that $P^0_{zz'} > 0$ if and only if $r(z, z') = 0$. In other words, transitions that can occur under P^0 have zero resistance. For convenience, we shall adopt the convention that $r(z, z') = \infty$ if $P^\varepsilon_{zz'} = P^0_{zz'} = 0$ for all $\varepsilon \in [0, \varepsilon^*]$, so that $r(z, z')$ is defined for all ordered pairs (z, z').

Since P^ε is irreducible for each $\varepsilon > 0$, it has a unique stationary distribution, which we shall denote by μ^ε. A state z is *stochastically stable* (Young, 1993a) if

$$\lim_{\varepsilon \to 0} \mu^\varepsilon(z) > 0. \tag{3.13}$$

As we shall show in theorem 3.1 below, $\lim_{\varepsilon \to 0} \mu^\varepsilon(z) = \mu^0(z)$ exists for every z, and the limiting distribution μ^0 is a stationary distribution of P^0. It follows in particular that every regular perturbed Markov process has at least one stochastically stable state. Intuitively, these are the states that are most likely to be observed over the long run when the random perturbations are small. In a moment we shall show how to compute the stochastically stable states using a suitably defined potential function. First, however, let us observe that adaptive play with parameters m, s, ε is a regular perturbed Markov process. Given an n-person game G with finite strategy space $X = \prod X_i$, the process operates on the finite space X^m of length-m histories. Given a history $h = (x^1, x^2, \ldots, x^m)$ at time t, the process moves in the next period to a state of form $h' = (x^2, x^3, \ldots, x^m, x)$ for some $x \in X$. Any such state h' is said to be a *successor* of h. Before choosing an action, an i-player draws a sample of size s from the m previous choices in h of each class $j \neq i$, the samples being independent among distinct classes j. The action x_i is an *idiosyncratic choice* or *error* if and only if there exists *no* set of $n - 1$ samples in h (one from each class $j \neq i$) such that x_i is a best reply to the product of the sample frequency distributions. Note that the concept of error depends on the preceding state h. For each successor h' of h, let $r(h, h')$ denote the total number of errors in the rightmost element of h'. Evidently $0 \leq r(h, h') \leq n$. It is easy to see that the probability of the transition $h \to h'$ is on the order of $\varepsilon^{r(h,h')}(1-\varepsilon)^{n-r(h,h')}$, where we omit a multiplicative constant that is independent of ε. If h' is not a successor of h, the probability of the transition $h \to h'$ is zero. Thus the process $P^{m,s,\varepsilon}$ approaches $P^{m,s,0}$ at a rate that is approximately exponential in ε; moreover it is irreducible whenever $\varepsilon > 0$. It follows that $P^{m,s,\varepsilon}$ is a regular perturbed Markov process.

We will now show how to compute the stochastically stable states for any regular perturbed Markov process P^ε on a finite state space Z. Let P^0 have recurrent classes $E_1, E_2, \ldots, E_K$. For each pair of distinct recurrent classes E_i and E_j, $i \neq j$, an *ij-path* is a sequence of states $\zeta = (z_1, z_2, \ldots, z_q)$ that begins in E_i and ends in E_j. The *resistance* of this path is the sum of the resistances of its edges, that is, $r(\zeta) = r(z_1, z_2) + r(z_2, z_3) + \cdots + r(z_{q-1}, z_q)$. Let $r_{ij} = \min r(\zeta)$ be the least resistance over all ij-paths ζ. Note that by the definition of a recurrent class, r_{ij} must be positive, because there is no path of zero resistance from E_i to E_j.

Now construct a complete directed graph with K vertices, one for each recurrent class. The vertex corresponding to class E_j will be called j. The weight on the directed edge $i \to j$ is r_{ij}. Figure 3.3 shows an illustrative example with three classes. A tree rooted at vertex j (a *j-tree*) is a set of $K - 1$ directed edges such that, from every vertex different from j,

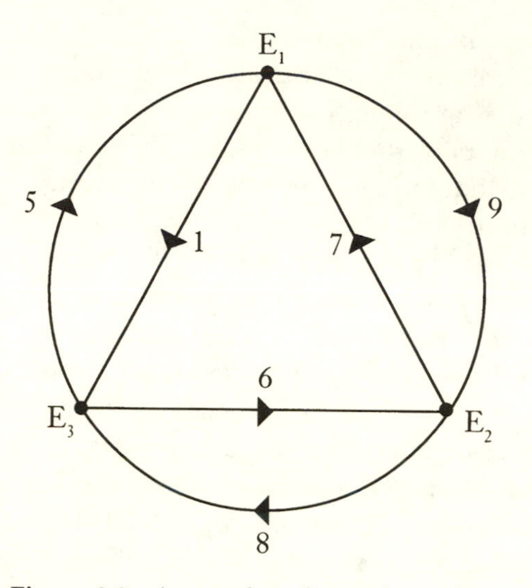

Figure 3.3. A complete directed graph on three vertices with resistances on edges.

there is a unique directed path in the tree to j. The example shown in Figure 3.3 has three j-trees rooted at each vertex j, and nine rooted trees altogether (see Figure 3.4).

The *resistance* of a rooted tree T is the sum of the resistances r_{ij} on the $K-1$ edges that compose it. The *stochastic potential* γ_j of the recurrent set E_j is defined as the minimum resistance over all trees rooted at j. Intuitively, when the noise parameter ε is small and positive, the process is most likely to follow paths that lead toward the recurrent classes having minimum potential. This suggests that the stochastically stable states are precisely those that sit in the classes having minimum potential. Formally this result can be stated as follows.[5]

THEOREM 3.1 (Young, 1993a). *Let P^{ε} be a regular perturbed Markov process, and let μ^{ε} be the unique stationary distribution of P^{ε} for each $\varepsilon > 0$. Then $\lim_{\varepsilon \to 0} \mu^{\varepsilon} = \mu^0$ exists, and μ^0 is a stationary distribution of P^0. The stochastically stable states are precisely those states that are contained in the recurrent class(es) of P^0 having minimum stochastic potential.*

We illustrate this result for the typewriter game with learning parameters $m = 4$ and $s = 2$. We have already seen that the unperturbed process (with $\varepsilon = 0$) has exactly two recurrent classes, each consisting of

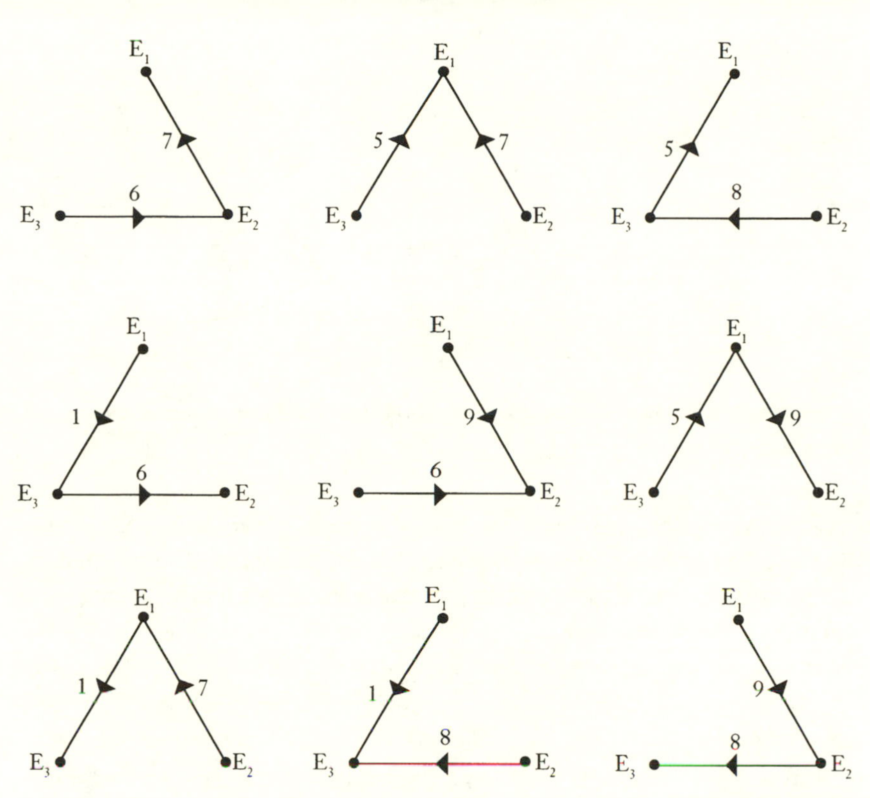

Figure 3.4. The nine rooted trees corresponding to the graph in Figure 3.3.

a single absorbing state: $E_1 = \{\text{all-D}\}$ and $E_2 = \{\text{all-Q}\}$. We need to find the least resistant path from E_1 to E_2, and also from E_2 to E_1. Consider the first case. If the process is in the all-D state, and exactly one player makes an error (plays Q), then we obtain a state such as the following:

DDDQ
DDDD

In the next period, both players will choose D as a best reply to any sample of size two; hence, if there are no further errors, the process will revert back (after four periods) to the state, all-D.

Suppose now that two players simultaneously make an error, so that we obtain a state of the following form:

DDDQ
DDDQ

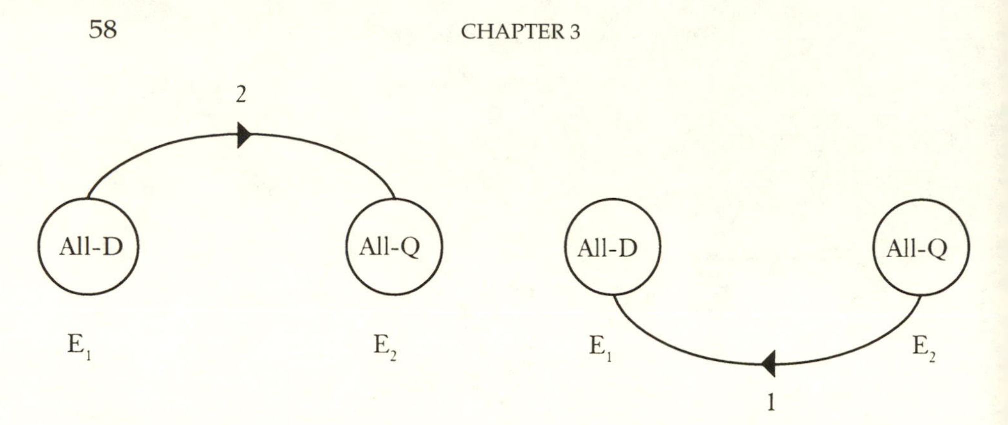

Figure 3.5. The resistances of the two rooted trees in the typewriter game with learning parameters $m = 4$, $s = 2$.

Again, the unique best reply to every subset of size two is D. Hence, if there are no more errors, the process drifts back to the all-D state. Suppose however that two errors occur in succession in the same population:

DDQQ
DDDD

From this state, there is a positive probability (though not a certainty) of transiting to the all-Q state with no further errors. The most probable transition from all-D to the above state occurs in two steps, each step having probability on the order of ε. Thus the resistance to moving from E_1 to E_2 is $r_{12} = 2$. On the other hand, if we begin in $E_2 = \{\text{all-Q}\}$, a single choice of D suffices for the process to evolve (with no further errors) to the all-*D* state. Hence, $r_{21} = 1$.

The two rooted trees for this example are quite trivial: each contains one edge, as shown in Figure 3.5. The tree with least resistance is the one that points from vertex 2 (all-Q) to vertex 1 (all-D). From the theorem, we therefore conclude that the unique stochastically stable state is all-D. In other words, when ε is small, the long-run probability of being in this state is far higher than that of being in any other state.

To get a feel for the extent to which the probability accumulates on all-D as a function of ε, we may simulate the process by Monte Carlo methods. (This is only practical when the state space is small, as in the present example.) The results for selected values of ε are summarized in Table 3.1. Note that, even with a sizable error rate (say, $\varepsilon = .20$), the

TABLE 3.1
Long-Run Probabilities of QWERTY and DVORAK Proportions in the Typewriter Game.

Proportion of D-players	ε .20	.10	.05	.01
1	.419	.659	.815	.961
7/8	.373	.277	.167	.039
6/8	.150	.053	.016	.001
5/8	.043	.009	—	—
4/8	.011	.001	—	—
3/8	.002	—	—	—
2/8	.001	—	—	—
1/8	—	—	—	—
0	—	—	—	—

Notes: Learning parameters are $m = 4$, $s = 2$. Estimated to three decimal places by Monte Carlo simulations. A dash indicates less than .0005 probability.

expected proportion of D-players is over 80 percent. This illustrates the point that the stationary distribution can put high probability on the stochastically stable states even when ε is fairly large.

The virtue of theorem 3.1 is that it tells us where the long-run probability of the learning process is concentrated when we lack a precise estimate of ε, but know that it is "small." If we did know ε precisely, we could (in theory) compute the actual distribution μ^{ε}. However, it would be very cumbersome to solve the stationarity equations (3.7) directly, because of the unwieldy size of the state space. Fortunately, there is a second way of computing μ^{ε} that yields analytically tractable results in certain situations.

Like the computation of the stochastic potential function, the computation of μ^{ε} is based on the notion of rooted trees. Unlike the former case, however, these trees must be constructed on the whole state space Z. Hence the approach is analytically useful only in certain special cases. Let P be any irreducible Markov process defined on a finite state space Z. (Note that P does not have to be a regular perturbed process.) Consider a directed graph having vertex set Z. The edges of this graph form a z-tree (for some particular $z \in Z$) if it consists of $|Z| - 1$ edges and from every vertex $z' \neq z$ there is a unique directed path from z' to z. Directed

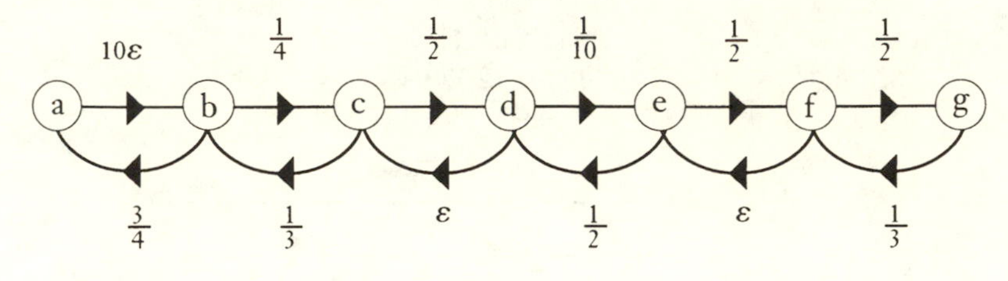

Figure 3.6. A Markov chain with a linear transition structure.

edges can be represented by ordered pairs of vertices (z, z'), and we can represent a z-tree T as a subset of ordered pairs. Let $\mathcal{T}_z$ be the family of all z-trees. Define the *likelihood* of z-tree $T \in \mathcal{T}_z$ to be

$$P(T) = \prod_{(z,z')\in T} P_{zz'}.$$

The following result is proved in the Appendix.

LEMMA 3.1 (Freidlin and Wentzell, 1984). *Let P be an irreducible Markov process on a finite state space Z. Its stationary distribution μ has the property that the probability $\mu(z)$ of each state z is proportional to the sum of the likelihoods of its z-trees, that is,*

$$\mu(z) = \nu(z)/\sum_{w\in Z} \nu(w), \text{ where } \nu(z) = \sum_{T\in\mathcal{T}_z} P(T).$$

To illustrate the difference between this result and theorem 3.1, consider a Markov process with seven states $Z = \{a, b, c, d, e, f, g\}$ and the transition probabilities shown in Figure 3.6. (It is understood that the probability of staying in a given state is one minus the total probability of leaving it in each period.)

This process has an essentially linear structure, that is, transitions can occur only to a "neighboring" state. This means that for each state z, there is a unique z-tree having nonzero probability. For example, the unique c-tree consists of the directed edges $a \to b$, $b \to c$, $d \to c$, $e \to d$, $f \to e$, $g \to f$. Lemma 3.1 tells us that the likelihood of a z-tree is proportional to the probability of z in the stationary distribution. It is a simple matter to compute the likelihood of each such tree—the tree's likelihood is just the product of the probabilities on its edges. A

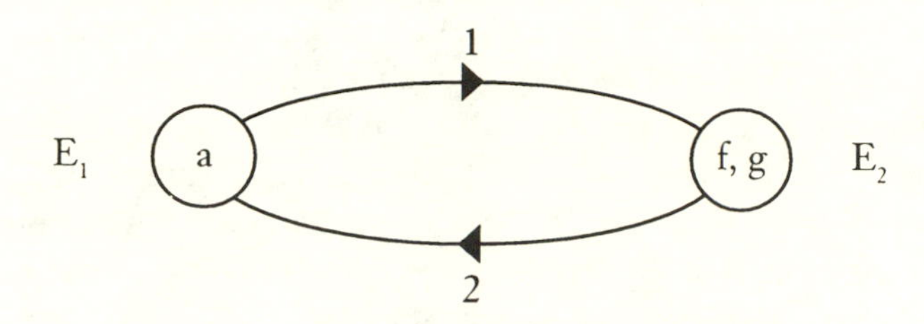

Figure 3.7. Resistances between the two recurrent classes in Markov chain shown in Figure 3.6.

little computation shows that the likelihoods of the various z-trees, and therefore the relative probabilities of the various states z are as follows:

State:	a	b	c	d	e	f	g
Relative probability:	$\varepsilon^2/24$	$5\varepsilon^3/9$	$5\varepsilon^3/12$	$5\varepsilon^2/24$	$\varepsilon^2/24$	$\varepsilon/48$	$\varepsilon/32$

Note that this estimate is *exact* for each $\varepsilon > 0$; it is not a limiting result as ε approaches zero. When ε is small, however, we can see immediately that states f and g will be much more likely than any of the other states.

We will now show how theorem 3.1 leads to the same conclusion by a different route. Let P^ε denote the transition matrix of the process as a function of ε. First, we observe that the unperturbed process P^0 has exactly two recurrent classes: $E_1 = \{a\}$, and $E_2 = \{f, g\}$. Then we compute the resistances from E_1 to E_2, and vice versa (see Figure 3.7). On one hand, the least resistant path (among all feasible paths) from E_1 to E_2 has probability $10\varepsilon \times 1/4 \times 1/2 \times 1/10 \times 1/2 \times 1/2 = \varepsilon/32$. Since this involves ε to the first power, the resistance is $r_{12} = 1$. On the other hand, the least resistant path from E_2 to E_1 involves ε to the second power; hence $r_{21} = 2$. The graph from which we compute the stochastic potential is therefore quite trivial: it consists of two vertices and two directed edges. The least resistant E_1-tree consists of the single edge $E_2 \to E_1$, whose resistance is 2. The least resistant E_2-tree consists of the single edge $E_1 \to E_2$, whose resistance is 1. Thus the stochastic potential of E_1 equals 2, and the stochastic potential of E_2 equals 1. It follows from theorem 3.1 that E_2 is stochastically stable, which is exactly what the previous calculation showed in a more detailed way.

While lemma 3.1 gives a more precise estimate of the stationary distribution, both approaches can be used to compute the stochastically stable states for any regular perturbed Markov process. The key difference is the amount of calculation needed to reach the desired conclusion. When the state space is large (and typically it will be very large), lemma 3.1 is impractical to apply because each state z will be associated with an astronomical number of z-trees. Theorem 3.1 provides a shortcut: to compute the stochastically stable states, we need only compute the resistance between each pair of recurrent classes of the unperturbed process. Since the number of such classes is often quite small, even when the state space is large, and computing least resistant paths is usually straightforward, the computational burden is greatly reduced.

In subsequent chapters, we shall examine the implications of these ideas for various classes of games. The theory provides sharp predictions about the equilibrium (and disequilibrium) configurations that are selected by simple learning rules. Moreover, it shows how key solution concepts in the game theory literature that are traditionally justified by high-rationality arguments—including the Nash bargaining solution in bargaining games (see Chapter 8), and efficient coordination equilibria in pure coordination games (see Chapter 9)—can emerge in low-rationality environments as well. It should be emphasized, however, that this approach is very general and can be used to study a variety of dynamic processes, not just the adaptive playing of games.

3.5 The Neighborhood Segregation Model

To illustrate this point, consider the neighborhood segregation model discussed in Chapter 1. Individuals are located at n possible positions around a circle, and they are of two types: A and B. A *state* z of the system is an assignment of A or B to each position. To avoid trivialities we assume that there are at least two individuals of each type. Individuals are *discontent* if both immediate neighbors are of the opposite type; otherwise they are *content*.

In each period, one pair of individuals is selected at random, where all pairs are equally likely to be chosen. Consider such a pair of individuals, say i and j. The probability that they trade depends on their prospective *gains* from trade. Let us assume that every trade involves moving costs. Thus there can be positive gains from trade only if the partners are of opposite types and at least one of them (say i) was discontent before and is content afterwards. This means that before the trade i was surrounded by people of the opposite type, so in fact both i and

j are content afterwards. (We shall assume that if j was content before and after, i can compensate j for his moving costs and still leave both better off.) Such Pareto improving trades are said to be *advantageous*; all other trades are *disadvantageous*.

Assume that each advantageous trade occurs with high probability, and that each disadvantageous trade occurs with a low probability that decreases exponentially with the partners' net loss in utility. Specifically, let us suppose that there exist real numbers $0 < a < b < c$ such that the probability of a disadvantageous trade is ε^a if one partner's increase in contentment is offset by the other's decrease in contentment (so the net losses involve only moving costs), the probability is ε^b if both partners were content before and one is discontent after, and it is ε^c if both were content before and both are discontent after. (These are the only possibilities.) Advantageous trades, by contrast, occur with probabilities that approach one as $\varepsilon \to 0$; beyond this we need not specify the probabilities exactly. The resulting Markov process P^ε is irreducible for every $\varepsilon > 0$, and satisfies conditions (3.11) and (3.12).

A state of the unperturbed process P^0 is absorbing if each person is located next to somebody of his or her own type. A state is *segregated* if the As form one contiguous group, and the Bs form the complementary group. In Chapter 1 we asserted that the segregated states are the only stochastically stable states of the perturbed process P^ε. We shall now establish this fact using the theory described in the preceding section.

The first step is to identify the recurrent communication classes of P^0. These obviously include the absorbing states, which are singleton recurrent classes. To prove that these are the only recurrent classes, consider a state that is not absorbing. It contains at least one discontent individual, say i, who we may assume is of type A. Going clockwise around the circle, let i' be the next individual of type A. (Recall that there are at least two individuals of each type.) The individual just before i' must be of type B. Call this individual j. If i and j trade places, both will be content afterwards. In any given period there is a positive probability that this pair will in fact be drawn, and that they will trade. The resulting state has fewer discontent individuals. Continuing in this manner, we see that from any nonabsorbing state there is a positive probability of transiting to an absorbing state within a finite number of periods. Hence the absorbing states are the only recurrent states.

Denote the set of absorbing states by Z^0. For any two states z and z' in Z^0, let $r(z, z')$ denote the *least* resistance among all paths from z to z'. The stochastic potential of $z \in Z^0$ is defined to be the resistance of the

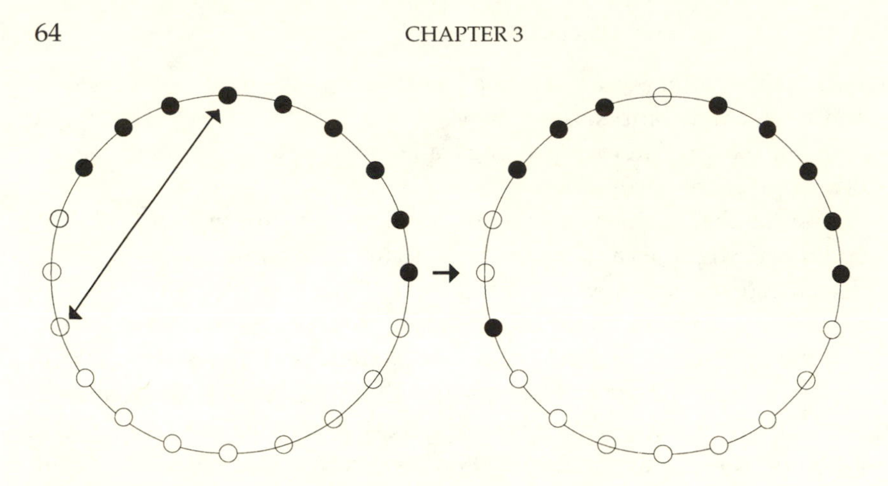

Figure 3.8. Single disadvantageous trade in segregated state resulting in state with isolated individuals.

minimum resistance z-tree on the set of nodes Z^0. By theorem 3.1, the stochastically stable states are those with minimum stochastic potential. It therefore suffices to show that an absorbing state has minimum potential if and only if it is segregated.

To prove this we proceed as follows: let $Z^0 = Z^s \cup Z^{ns}$ where Z^s is the set of segregated absorbing states and Z^{ns} is the set of nonsegregated absorbing states. We claim that: (i) for every $z \in Z^{ns}$, every z-tree has at least one edge with resistance b or c (which are greater than a); and (ii) for every $z \in Z^s$, there exists a z-tree in which every edge has resistance exactly equal to a.

Assume for the moment that (i) and (ii) have been established. In any z-tree there are exactly $|Z^0| - 1$ edges, and the resistance of each edge is *at least a*. It follows from (i) and (ii) that the stochastic potential of every segregated state equals $a|Z^0| - a$, while the stochastic potential of every nonsegregated state is at least $a|Z^0| - 2a + b$, which is strictly larger. Theorem 3.1 therefore implies that the segregated states are precisely the stochastically stable states.

To establish (i), let $z \in Z^{ns}$ be a nonsegregated absorbing state. Given any z-tree T, there exists at least one edge in T that is directed from a segregated absorbing state z^s to a nonsegregated absorbing state z^{ns}. We claim that any such edge has a resistance of at least b. The reason is that any trade that breaks up a segregated state must create at least one discontent individual, hence the probability of such a trade is either ε^b or ε^c (see Figure 3.8). Thus the resistance of the edge from z^s to z^{ns} must be at least b, which establishes (i). To establish (ii), let $z \in Z^s$ be a segregated absorbing state. From each state $z' \neq z$ we shall construct a sequence of absorbing states $z' = z^1, z^2, \ldots, z^k = z$ such that $r(z^{j-1}, z^j) = a$ for $1 < j \leq k$. Call this a $z'z$-path. We shall carry out the construction so that the union of all of the directed edges on all of these paths forms a z-tree. Since each edge has a resistance of a and the tree has $|Z^0| - 1$ edges, the total resistance of the tree is $a|Z^0| - a$ as claimed in (ii).

Suppose first that z' is also segregated, that is, z' consists of a single contiguous A-group and a complementary contiguous B-group. Label the positions on the circle $1, 2, \ldots, n$ in the clockwise sense. Let the first member of the A-group trade places with the first member of the B-group. Since both were content before and after, this trade has probability ε^a. It also results in a new absorbing state, which shifts the A-group and the B-group by one position clockwise around the circle. Hence within n steps we can reach any absorbing state, and in particular we can reach z. Thus we have constructed a sequence of absorbing states that leads from z' to z, where the resistance of each successive pair in the sequence equals a.

Suppose alternatively that z' is not segregated. Moving clockwise from position 1, let **A** denote the first complete group of contiguous As. Let **B** be the next group of Bs, and **A**$'$ the next group of As. Since z' is absorbing, each of these groups contains at least two members. Let the first player in **A** trade places with the first player in **B**. Since both players were content before and after the trade, its probability equals ε^a. This trade shifts group **A** one position clockwise and reduces by one the number of B players between **A** and **A**$'$. It either results in a new absorbing state, or else a single B player remains between **A** and **A**$'$. In the latter case this B player can then trade with the first player in group **A**, and this trade has zero resistance. The result is an absorbing state with fewer distinct groups of As and Bs.

Repeat the process described in the preceding paragraph until all the As are contiguous and all the Bs are contiguous. Then continue as in the earlier part of the argument until we reach the target state z. This construction yields a sequence of absorbing states that begins at z' and ends at z, where the resistance between each successive pair of states is a. The path contains no cycles because the number of distinct groups never increases; indeed with each transition one of the groups shrinks until it is eliminated. Thus the union of these paths forms a z-tree whose total resistance is $a|Z^0| - a$. This concludes the proof that the stochastically stable states are precisely the segregated ones.[6]

Chapter 4

ADAPTIVE LEARNING IN SMALL GAMES

IN THIS CHAPTER, we explore the properties of adaptive learning in two-person games, where each player has exactly two actions. Although this is a special case, it can be used to illustrate a wide variety of social and economic interactions. We begin by introducing the classical notion of risk dominance in 2×2 games, and then show that it coincides with the stochastically stable outcome under a variety of assumptions about the learning process. In larger games, however, the two concepts differ, as we shall show in Chapter 7.

4.1 RISK DOMINANCE

Consider a two-person game G with payoff matrix

	1	2
1	a_{11}, b_{11}	a_{12}, b_{12}
2	a_{21}, b_{21}	a_{22}, b_{22}

(4.1)

G is a coordination game with pure strategy Nash equilibria (1, 1) and (2, 2) if and only if the payoffs satisfy the inequalities

$$a_{11} > a_{21}, b_{11} > b_{12}, a_{22} > a_{12}, b_{22} > b_{21}. \qquad (4.2)$$

Equilibrium (1, 1) is *risk dominant* if

$$(a_{11} - a_{21})(b_{11} - b_{12}) \geq (a_{22} - a_{12})(b_{22} - b_{21}). \qquad (4.3)$$

Similarly, equilibrium (2, 2) is risk dominant if the reverse inequality holds (Harsanyi and Selten, 1988). If the inequality holds strictly, the corresponding equilibrium is *strictly* risk dominant.[1]

Risk dominance has a particularly simple interpretation when the game is symmetric. Consider the symmetric payoff matrix

	1	2
1	a, a	c, d
2	d, c	b, b

$$a > d, b > c. \tag{4.4}$$

Imagine that each player is uncertain about the action that the other player is going to take. If each has a flat prior (attaches a fifty-fifty probability to the intended action of the other), the expected payoff from playing action 1 is $(a+c)/2$, while the expected payoff from action 2 is $(d+b)/2$. An expected utility maximizer therefore chooses action 1 only if $(a+c)/2 \geq (d+b)/2$, that is, only if $a-d \geq b-c$. This is equivalent to saying that action 1 is risk dominant. In other words, when each player assigns equal odds to his or her opponent's playing action 1 or 2, both players will choose a risk-dominant action.

In the more general case (4.1), we can motivate risk dominance as follows. Define the *risk factor* of equilibrium (i, i), where $i = 1, 2$, to be the smallest probability p such that if one player believes the other player is going to play action i with probability strictly greater than p, then i is the unique optimal action to take.[2] Consider, for example, equilibrium $(2, 2)$. For the row player, the smallest such p satisfies

$$a_{11}(1-p) + a_{12}p = a_{21}(1-p) + a_{22}p,$$

that is,

$$p = (a_{11} - a_{21})/(a_{11} - a_{12} - a_{21} + a_{22}). \tag{4.5}$$

For the column player, the smallest such p satisfies

$$b_{11}(1-p) + b_{21}p = b_{12}(1-p) + b_{22}p,$$

that is,

$$p = (b_{11} - b_{12})/(b_{11} - b_{12} - b_{21} + b_{22}). \tag{4.6}$$

In general, let

$$\begin{aligned} \alpha &= (a_{11} - a_{21})/(a_{11} - a_{12} - a_{21} + a_{22}) \text{ and} \\ \beta &= (b_{11} - b_{12})/(b_{11} - b_{12} - b_{21} + b_{22}). \end{aligned} \tag{4.7}$$

The risk factor for equilibrium (2, 2) is therefore $\alpha \wedge \beta$, where in general $x \wedge y$ denotes the minimum of x and y. Similarly, the risk factor for equilibrium (1, 1) is $(1-\alpha) \wedge (1-\beta)$. A risk-dominant equilibrium in a 2×2 game is an equilibrium whose risk factor is lowest. Thus equilibrium (1, 1) is risk dominant if and only if

$$\alpha \wedge \beta \geq (1-\alpha) \wedge (1-\beta). \tag{4.8}$$

Equilibrium (2, 2) is risk dominant if and only if the reverse inequality holds. A little algebraic manipulation shows that (4.8) is equivalent to

$$(a_{11}-a_{21})(b_{11}-b_{12}) \geq (a_{22}-a_{12})(b_{22}-b_{12}). \tag{4.9}$$

In other words, in a 2×2 game an equilibrium is risk dominant if and only if it maximizes the product of the gains from unilateral deviation (Harsanyi and Selten, 1988).

4.2 Stochastic Stability and Risk Dominance in 2 × 2 Games

Recall that a *convention* is a state of form $(x, x, \ldots, x)$ where $x = (x_1, x_2, \ldots, x_n)$ is a strict Nash equilibrium of G. In such a state, everyone will continue to play their part in x (barring errors), because x_i is the unique best response given i's expectations that everyone else will play their part in x.

THEOREM 4.1. *Let G be a 2×2 coordination game, and let $P^{m,s,\varepsilon}$ be adaptive learning with memory m, sample size s, and error rate ε.*

(i) *If information is sufficiently incomplete ($s/m \leq 1/2$), then from any initial state, the unperturbed process $P^{m,s,0}$ converges with probability one to a convention and locks in.*
(ii) *If information is sufficiently incomplete ($s/m \leq 1/2$), and s and m are sufficiently large, the stochastically stable states of the perturbed process correspond one to one with the risk-dominant conventions.*

PROOF. Let G be a 2×2 coordination game with payoff matrix (4.1) satisfying the inequalities (4.2), so that both (1, 1) and (2, 2) are strict Nash equilibria. Let h_1 and h_2 denote the corresponding conventions for some fixed value of m. The *basin of attraction* of state h_i is defined to be the set of states h such that there is a positive probability of moving in a finite number of periods from h to h_i under the unperturbed process P^0. Let $\mathcal{B}_i$ denote the basin of attraction of h_i, $i = 1, 2$. To prove statement (i) of the theorem, we need to show that $\mathcal{B}_1 \cup \mathcal{B}_2$ covers the state space.

Let $h = (x^{t-m+1}, \ldots, x^t)$ be an arbitrary state. There is a positive probability that both players sample the particular set of precedents $x^{t-s+1}, \ldots, x^t$ in every period from $t + 1$ to $t + s$ inclusive. Since $\varepsilon = 0$, each of them plays a best reply. Assume for the moment that these best replies are unique, say, $(x_1^*, x_2^*) = x^*$. Then we obtain a *run* $(x^*, x^*, \ldots, x^*)$ from period $t + 1$ to period $t + s$. (If there are ties in best reply for some player, there is still a positive probability that the *same* pair of best replies $(x_1^*, x_2^*) = x^*$ will be chosen for s periods, because all best replies have a positive probability of being chosen.) Note that this argument uses the assumption that $s \leq m/2$. If s is too large relative to m, some of the precedents $x^{t-s+1}, \ldots, x^t$ will have "died out" by period $t + s$, which would contradict our assumption that they sample a *fixed* set of precedents for s periods in succession.

Suppose, on the one hand, that x^* is a coordination equilibrium, that is, $x^* = (1, 1)$ or $x^* = (2, 2)$. There is a positive probability that, from period $t + s + 1$ through period $t + m$ both players will sample from the run. The unique best reply by i to any such sample is x_i^* $(i = 1, 2)$. Hence, by the end of period $t + m$, there is a positive probability that the process will have reached the convention $(x^*, x^*, \ldots, x^*)$, and we are done.

Suppose, on the other hand, that x^* is not a coordination equilibrium. Then $x^* = (1, 2)$ or $x^* = (2, 1)$. Without loss of generality assume that $x^* = (1, 2)$. There is a positive probability that, from period $t + s + 1$ through period $t + 2s$, the row player will continue to sample from the sequence $(x^{t-s+1}, \ldots, x^t)$ and play 1 as best reply. There is also a positive probability that, simultaneously, the column player will sample from the run and therefore play 1 as best reply. Thus, from period $t + s + 1$ through period $t + 2s$, we obtain a run of form $(1, 1), (1, 1), \ldots, (1, 1)$. From this point, it is clear that, with positive probability, the process converges to the convention h_1.

Thus we have shown that from any initial state, there is a positive probability of reaching h_1 and/or h_2 in a finite number of periods. It follows that the only recurrent classes of the unperturbed process are the absorbing states h_1 and h_2, as claimed in statement (i).

To establish statement (ii), we apply theorem 3.1. Let r_{12}^s denote the least resistance among all paths that begin at h_1 and end at h_2 as a function of the sample size s. Clearly this is the same as the least resistance among all paths that begin at h_1 and end in $\mathcal{B}_2$, because after entering $\mathcal{B}_2$, no further errors are needed to reach h_2. Define r_{21}^s similarly.

Let α and β be defined as in (4.7). Also, let $\lceil y \rceil$ denote the least integer greater than or equal to y for any real number y. Suppose that the process is in absorbing state h_1, where both the row player and the column player have chosen action 1 for m periods in succession. For a row player to prefer action 2 over action 1, there must be at least $\lceil \alpha s \rceil$ instances of action 2 in the row player's sample. This will happen with positive probability if a succession of $\lceil \alpha s \rceil$ column players choose action 2 by mistake. (Note that this uses the assumption that all samples are drawn with positive probability.) The probability of these events is on the order of $\varepsilon^{\lceil \alpha s \rceil}$. Similarly, a column player will prefer action 2 over action 1 only if there are at least $\lceil \beta s \rceil$ instances of action 2 in the column player's sample. This will happen with positive probability if a succession of $\lceil \beta s \rceil$ column players choose action 2 by mistake, which has a probability on the order of $\varepsilon^{\lceil \beta s \rceil}$.

It follows that the resistance to going from h_1 to h_2 is $r_{12}^s = \lceil \alpha s \rceil \wedge \lceil \beta s \rceil$. A similar calculation shows that $r_{21}^s = \lceil (1-\alpha)s \rceil \wedge \lceil (1-\beta)s \rceil$. By theorem 3.1, h_1 is stochastically stable if and only if $r_{12}^s \geq r_{21}^s$; similarly h_2 is stochastically stable if and only if $r_{12}^s \leq r_{21}^s$. If one equilibrium is strictly risk dominant, say, equilibrium (1, 1), then $r_{12}^s > r_{21}^s$ for all sufficiently large s, so the corresponding convention is stochastically stable. Suppose, on the other hand, that the two equilibria are tied for risk dominance. Then $\alpha = 1 - \beta$ and $r_{12}^s = r_{21}^s$ for all s, so both h_1 and h_2 are stochastically stable. This concludes the proof of theorem 4.1.

We do not claim that the bound on incompleteness $s/m \leq 1/2$ is the best possible, but some degree of incompleteness is necessary for part (i) of theorem 4.1 to hold. To see why, consider the etiquette game described in Chapter 2. Let $s = m$ and suppose that the process starts out in a miscoordinated state where either both players always yielded for s periods or they always failed to yield. Because the players react to the full history, and the error rate is zero, they are sure to miscoordinate again. This miscoordination continues forever, and the process never reaches an absorbing state. Statement (i) of the theorem asserts that such cycling behavior cannot happen when information is sufficiently incomplete, because incomplete sampling provides enough random variation (even without errors) to shake the process out of a cycle.

4.3 Who Goes First?

To illustrate this result, consider the general game of who goes first. The etiquette game is one example; another with greater consequence is the proposal game. Is the man expected to propose to the woman, or the other way around? This matter is obviously influenced by social custom: both sides have expectations about who should take the initiative, and these expectations are shaped by what other people have done in similar circumstances. If the person who is supposed to move first does not do so, the other may take it as a sign of disinterest. If the person who is not supposed to move first does so anyway, the other may view it as presumptuous. In short, making the wrong move can have serious consequences.

The implicit game here is to coordinate on the rules of another game (who goes first, who goes second). We may view this metagame as one of pure coordination: if the parties fail to coordinate, their payoffs are zero; if they coordinate they receive higher payoffs than if they miscoordinate. To make the example more interesting, we may suppose that the payoffs to the two sides are asymmetric. For the sake of concreteness, let the payoffs be as follows:

The Proposal Game

		Men	
		Respond	Propose
Women	Propose	10, 7	0, 0
	Respond	0, 0	9, 10

The stochastically stable equilibrium is the one that maximizes the product of the parties' payoffs. In the present example, this is the equilibrium (9, 10) in which the men propose and the women respond. In other words, under repeated interactions by many myopic agents, the "men propose" equilibrium will emerge as the standard, or "conventional," one most of the time.

While this example is highly stylized and the payoffs are invented merely for purposes of illustration, the general point is that *the stability of a convention depends on its welfare consequences for individuals*. Furthermore, the choice of convention does not occur at the individual level, but emerges as an unintended consequence of many individuals responding to their immediate environment. This example also illustrates that games are not always given a priori, as game theorists like to assume; rather, the rules of the game themselves are social constructs (conventions) that are governed by evolutionary forces. To play a game, people must have common expectations about what the rules of the game are, and it seems reasonable to assume that these expectations are shaped (to some degree) by precedent. The theory therefore suggests that the rules of conventional games depend on their expected payoffs, and that when the competition is between two alternative game forms, the one that maximizes the product of the parties' expected payoffs is the more likely to be observed in the long run.

4.4 Playing the Field

A natural variant of the learning model arises when individuals constitute a single population and they are matched to play a *symmetric* game. This is essentially the case considered by Kandori, Mailath, and Rob (1993). Consider, for example, the currency game described in Chapter 1. At the beginning of each period, a person is drawn at random and he or she decides whether to carry gold or silver for all transactions that occur within that period. The expected payoff depends on the relative proportions of people carrying gold and people carrying silver in the general population. For the sake of concreteness, we shall suppose the following payoffs:

The Currency Game

	Gold	Silver
Gold	3, 3	0, 0
Silver	0, 0	2, 2

Thus, if p is the proportion holding gold in the general population, a person carrying gold has an expected payoff per period of $3p$, while a person holding silver has an expected payoff of $2(1-p)$.

In general, let G be a symmetric two-person game with strategy space X_0 that is played by a single population consisting of m individuals. The payoff functions are $u_1(x, x') = u_2(x', x)$ for the row player and column player, respectively. At the beginning of each period, each person resolves to play a given pure strategy against all comers during that period. For each $x \in X_0$, let k_x^t denote the number of individuals committed to playing strategy x in period t. The *state* at time t is therefore a vector $k^t = (k_x^t)$ of integers such that $\sum_x k_x^t = m$.

In this framework, adaptive learning works as follows. Let s be the sample size (an integer between 1 and m), and let $\varepsilon \in [0, 1]$ be the error rate. Assume that the state at the end of period t is k^t. At the beginning of period $t + 1$:

(i) One agent is drawn from the population at random.
(ii) With probability $(1 - \varepsilon)$, the agent draws a random sample of size s, without replacement, from the frequency distribution k^t, and plays a best reply to the resulting sample proportions $\hat{p}^t$. If there are ties in best reply each is played with equal probability.
(iii) With probability ε, the agent chooses an action in X_0 at random, each with equal probability.

This one-population version of adaptive learning is structurally similar to (though not identical with) the two-population process described earlier. In particular, if G is a symmetric 2×2 coordination game, then whenever $1 \leq s \leq m$ the unperturbed process converges with probability one to a convention, and the risk dominant conventions are stochastically stable whenever s and m are sufficiently large.

4.5 Computing the Stationary Distribution

Stochastic stability tells us what states survive with positive probability when the background noise is vanishingly small, but it does not say how probable these states are when the noise is merely "small." A priori, we cannot say how much noise is "reasonable," since this depends on the application at hand. Lacking such an estimate, it is of interest to ask how closely the stationary distribution μ^ε *approximates* its asymptotic limit when the noise level is positive. In this section, we show how to obtain a precise estimate of the distribution μ^ε for 2×2 playing-the-field games that gives a sense of how strongly the stochastically stable equilibrium is selected when the noise is small but not vanishingly small. The surprising answer is that the selection can be sharp even for sizable values of ε (e.g., $\varepsilon = .05$ or $.10$) so long as the population size is also large.[3] The reason for this will soon become apparent.

Consider the one-population learning model for the 2×2 symmetric coordination game in (4.4). We shall identify the state k^t at time t with the number of agents playing action 1; thus, the state space is one-

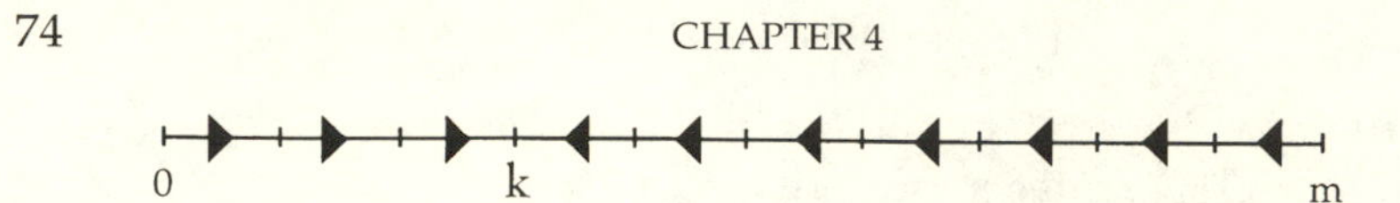

Figure 4.1. The unique k-tree having nonzero probability for a given state k.

dimensional. Let $(\gamma, 1-\gamma)$ be the mixed strategy equilibrium for each player, that is, $\gamma = (b-c)/(a-d+b-c)$. Assume that $\gamma < 1/2$, that is, equilibrium (1, 1) is strictly risk dominant. It will be convenient to assume that sampling is complete ($s = m$), that is, each player reacts to the whole distribution, including himself. Let $0 \leq k^t \leq m$ be the state in period t (the number of agents who played action 1). Assume that, with probability $1-\varepsilon$, the player selected in period $t+1$ chooses a best reply to the probability distribution $(k^t/m, 1-k^t/m)$, and that, with probability ε, he chooses strategy 1 or 2 at random, each with probability $\varepsilon/2$. Fix $\varepsilon \in (0, 1)$. Let P^m denote the transition matrix of this process; that is, $P^m_{kk'}$ is the probability of moving from state k to state k' in one period. Note that the process either stays in the same state or moves to an adjacent state: $P^m_{kk'} > 0$ only if $k' = k-1, k$, or $k+1$. The transition probabilities are as follows:

$$\begin{aligned} 0 \leq k < \gamma m,\ & P^m_{k,k+1} = (1-k/m)(\varepsilon/2),\ P^m_{k,k-1} = (k/m)(1-\varepsilon/2), \\ k = \gamma m,\ & P^m_{k,k+1} = (1-k/m)(1/2),\ P^m_{k,k-1} = (k/m)(1/2), \\ \gamma m < k \leq m,\ & P^m_{k,k+1} = (1-k/m)(1-\varepsilon/2),\ P^m_{k,k-1} = (k/m)(\varepsilon/2). \end{aligned} \tag{4.10}$$

Since the only feasible one-period transitions are to adjacent states (or the same state), each state k is associated with exactly one k-tree having non-zero probability, namely, the tree T_k in which all edges lie on the line and are directed toward k (see Figure 4.1).

Fix $\varepsilon \in (0, 1)$. For each positive integer m, let $\mu^m(k)$ denote the unique stationary distribution of the process P^m on the state space $0 \leq k \leq m$. By lemma 3.1, $\mu^m(k)$ is proportional to the product of the probabilities on the edges of the unique k-tree T_k. We claim that, when m is sufficiently large, $\mu^m(k)$ puts almost all the probability on states k such that k/m is close to $1-\varepsilon/2$. To establish this, let us extend (4.10) by defining the probabilities of right and left transitions for every real number $w \in [0, 1]$ as follows:

$$\begin{aligned} 0 \leq w < \gamma,\quad & R(w) = (1-w)(\varepsilon/2), && L(w) = w(1-\varepsilon/2), \\ w = \gamma,\quad & R(w) = (1-w)(1/2) && L(w) = w(1/2), \\ \gamma < w \leq 1,\quad & R(w) = (1-w)(1-\varepsilon/2), && L(w) = w(\varepsilon/2). \end{aligned} \tag{4.11}$$

For each $w \in [0, 1]$, define

$$\nu^m(w) = \prod_{i<wm} R(i/m) \prod_{i>wm} L(i/m). \tag{4.12}$$

where i ranges over the integers $0, 1, 2, \ldots, m$. Note that for integer k, $\nu^m(k/m)$ is just the product of the transition probabilities of the edges in the unique k-tree T_k, hence $\nu^m(k/m)$ is proportional to $\mu^m(k)$. We shall study the shape of $\nu^m(\cdot)$ when m becomes large. From (4.12) we have

$$(1/m) \ln \nu^m(w) = (1/m) \left[\sum_{i<wm} \ln R(i/m) + \sum_{i>wm} \ln L(i/m) \right]. \tag{4.13}$$

For every $w \in [0, 1]$, define the function $V(w)$ as follows:

$$V(w) = \int_0^w \ln R(u)\, du + \int_w^1 \ln L(u)\, du. \tag{4.14}$$

Then

$$\lim_{m \to \infty} (1/m) \ln \nu^m(w) = V(w), \tag{4.15}$$

and the convergence is uniform on $[0, 1]$. In effect, $-V(w)$ is the stochastic potential of the state w when the population size m tends to infinity and the error rate ε is fixed. (This construction is quite general, and works for a wide variety of one-dimensional processes.)

Let w^* be a point at which $V(w)$ achieves its maximum. The first-order condition is $V'(w^*) = 0$, which holds if and only if $R(w^*) = L(w^*)$. This has two solutions: $w^* = 1 - \varepsilon/2$ and $w^* = \varepsilon/2$. A direct evaluation of (4.14) shows that $w^* = 1 - \varepsilon/2$ is the unique global maximum. For every small $\delta > 0$, let $F_\delta = \{w\colon |w - w^*| \geq \delta\}$ and $N_{\delta/2} = \{w\colon |w - w^*| \leq \delta/2\}$. Then $\sup\{V(w)\colon w \in F_\delta\} < \inf\{V(w)\colon w \in N_{\delta/2}\}$, and

$$\lim_{m \to \infty} \left(\int_{F_\delta} e^{mV(w)}\, dw \Big/ \int_{N_{\delta/2}} e^{mV(w)}\, dw \right) = 0. \tag{4.16}$$

From this and (4.15) we conclude that $\nu^m(\cdot)$, and also $\mu^m(\cdot)$, are concentrated in a δ-neighborhood of $w^* = 1 - \varepsilon/2$ for all sufficiently large m, that is,

$$\lim_{m \to \infty} \mu^m(\{k\colon |k/m - (1 - \varepsilon/2)| \leq \delta\}) = 1. \tag{4.17}$$

We summarize this in the following result.

THEOREM 4.2. *Let G be a 2×2 symmetric coordination game with a strictly risk dominant equilibrium, and let $Q^{m,\varepsilon}$ be adaptive learning in the playing the field model with population size m, complete sampling, and error rate ε, $0 < \varepsilon < 1$. For every $\varepsilon' > \varepsilon$, the probability is arbitrarily high that at least $1 - \varepsilon'/2$ of the population is playing the risk dominant equilibrium when m is sufficiently large.*

This result shows that, even when individuals make independent errors at a substantial rate, the aggregation of these errors can result in fairly strong forces of selection at the population level.

Chapter 5

VARIATIONS ON THE LEARNING PROCESS

IN THIS CHAPTER we shall consider various extensions and alterations of the basic learning model. First, we shall ask what happens when people have different amounts of information and different utility functions, that is, when the populations are heterogeneous. This can affect the stochastically stable outcome, but it does not fundamentally alter the qualitative properties of the selection process. Second, we consider an alternative model of noise in which the probability of an error depends on the seriousness of the error, that is, on the loss in payoff it induces. We show that for symmetric 2×2 games this does not change the stochastically stable outcome, which remains the risk-dominant equilibrium. Third, we introduce an apparently minor modification of the model in which players attach equal weight to *all* past actions of their opponents (i.e., memory is infinite). As we shall see, this model has fundamentally different long-run properties from the model with finite memory. In particular its average behavior depends crucially on the initial conditions.

5.1 HETEROGENEITY IN INFORMATION

Consider the process by which agents gather information. So far, we have assumed that people have similar amounts of information (the same sample size), though, of course, they may not have the same information. Sample size can be viewed as an inherent characteristic of an agent, and reflects the fact that people acquire much of their information through existing social networks, rather than through a conscious search process. Thus the size of a person's sample is determined by the number of his contacts, which is a part of his endowment. What happens when people have different endowments, that is, when the populations are heterogeneous with respect to amount of information?

Let us first examine this issue by assuming that the row players have one (uniform) sample size s, and the column players have another (uniform) sample size s'. Let the error rate, ε, be the same for all players,

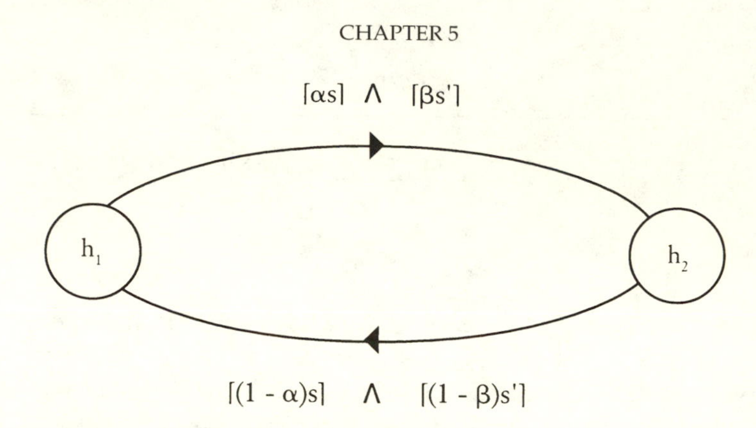

Figure 5.1. Resistances when row and column have sample sizes s and s', respectively.

and let memory, m, satisfy $m \geq 2s \vee 2s'$. (In general, $x \vee y$ denotes the maximum of x and y.) Let G be a 2×2 coordination game with payoffs

	1	2
1	a_{11}, b_{11}	a_{12}, b_{12}
2	a_{21}, b_{21}	a_{22}, b_{22}

(5.1)

$$a_{11} > a_{21}, b_{11} > b_{12}, a_{22} > a_{12}, b_{22} > b_{21}. \tag{5.2}$$

As in equation (4.7), define

$$\begin{aligned} \alpha &= (a_{11} - a_{21})/(a_{11} - a_{12} - a_{21} + a_{22}) \text{ and} \\ \beta &= (b_{11} - b_{12})/(b_{11} - b_{12} - b_{21} + b_{22}). \end{aligned} \tag{5.3}$$

As in the proof of theorem 4.1, it is straightforward to show that the resistance to transiting between the two absorbing states h_1 and h_2 is as shown in Figure 5.1.

Thus h_1 is stochastically stable if and only if

$$\lceil \alpha s \rceil \wedge \lceil \beta s' \rceil \geq \lceil (1-\alpha)s \rceil \wedge \lceil (1-\beta)s' \rceil. \tag{5.4}$$

The question now arises whether the class with more information is better off. The answer depends on the structure of the game. Consider,

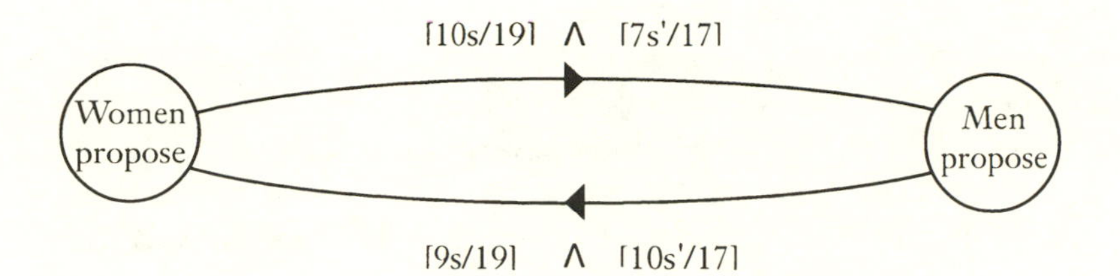

Figure 5.2. Resistances for the proposal game.

for example, the proposal game introduced in Section 4.3. Then $\alpha = 10/19$, $\beta = 7/17$, and the resistance diagram is as shown in Figure 5.2.

When $\lceil 9s/19 \rceil < \lceil 7s'/17 \rceil$ the "women propose" norm is stochastically stable, whereas if $\lceil 9s/19 \rceil > \lceil 7s'/17 \rceil$ the "men propose" norm is stochastically stable. In other words, when s'/s is sufficiently large (and s and m also are sufficiently large), selection favors the women: having *less* information is an advantage. The reason is that agents with less information are more flexible in responding to "weakness" by the other side, that is, quicker to pick up on and exploit errors that act in their favor.

This is only one side of the coin, however, because having less information also means that agents are more likely to give in to "strength" by the other side, that is, to errors that do not act in their favor. The balance between these two effects depends on the structure of the game. Consider, for example, a two-person interaction in which each side can adopt an aggressive or a defensive posture (the game of Chicken). The payoffs are as follows:

Game of Chicken

	Defensive	Aggressive
Aggressive	10, 5	0, 0
Defensive	7, 7	5, 10

Aggression by both is costly, but it is not an equilibrium for both to play defensively: in a pure equilibrium, one is aggressive and the other is defensive. Let the row population have sample size s and the column population sample size s'. It is straightforward to check that upper left is the stochastically stable outcome if $\lceil 3s'/8 \rceil < \lceil 3s/8 \rceil$, and lower right is the stochastically stable outcome if $\lceil 3s'/8 \rceil > \lceil 3s/8 \rceil$. Thus having more information is an advantage in this game.

This mode of analysis extends readily to situations where each population is heterogeneous. Let s be the smallest sample size among all members of the row population, and let s' be the smallest sample size among all members of the column population. It is straightforward to show that, if m is at least twice as large as $s \vee s'$, then the stochastically stable outcome is determined by the resistances shown in Figure 5.1.

5.2 HETEROGENEITY IN PAYOFFS

A similar approach can be applied when populations are heterogeneous with respect to both information and payoffs. Let each row agent r have sample size $s(r)$ and 2×2 payoff matrix $A(r) = [a_{ij}(r)]$, where the entries $a_{ij}(r)$ satisfy the conditions of (5.2). Similarly, let each column agent c have sample size $s(c)$ and 2×2 payoff matrix $B(c) = [b_{ij}(c)]$ satisfying (5.2). Thus the *structure* of the game is the same for any pair of agents, but they differ in their assessment of the relative utilities of the various outcomes. From each row agent's payoff matrix $A(r)$, we can compute $\alpha(r)$ as in (5.3); similarly, from each column agent's payoff matrix $B(c)$, we can compute $\beta(c)$. The *type* of an agent is a pair (s, γ), where s is the agent's sample size, and $\gamma = \alpha(r)$ if the agent r is drawn from the row population, and $\gamma = \beta(c)$ if the agent c is drawn from the column population. Let T be the set of all types represented in the row and column populations, and assume that T is *finite*. A simple extension of the proof of theorem 4.1 shows that convention h_1 is stochastically stable if and only if

$$\min_{(s,\gamma)\in T} \lceil \gamma s \rceil \geq \min_{(s,\gamma)\in T} \lceil (1-\gamma)s \rceil, \tag{5.5}$$

while convention h_2 is stochastically stable if and only if the reverse inequality holds.

5.3 ALTERNATIVE MODELS OF NOISE

We turn now to the question of how sensitive the stochastically stable outcome is to the way in which errors are modeled. Our assumption so far has been that the probability that a player makes an error is the same for all players and is independent of the state of the system. This *uniform error model* seems to be the most natural assumption in the absence of a specific account of why players make errors. If we allow an arbitrary error structure, in which the probability of choosing a nonbest reply is state-dependent, then as these error rates approach zero, the process can select *any* strict Nash equilibrium, depending on the details of the error structure (Bergin and Lipman, 1996).

We may equally well ask, however, whether we can relax the assumption of uniformity in plausible ways without changing the stochastically stable outcome. Here, we shall illustrate two such variations. Consider first the possibility that when the agents make errors, their choices are systematically biased in favor of one strategy or the other. To be concrete, let the game be as in (5.1) and assume that equilibrium (1, 1) is risk dominant. Suppose that each player is in the "error mode" with probability ε, and that when in the error mode, he or she chooses action 1 with probability λ and action 2 with probability $1-\lambda$, where $0 < \lambda < 1$. (In the previous model, we assumed that $\lambda = 1/2$.) When ε is small, the probabilities of transiting between the two conventions h_1 and h_2 are as follows: for some positive constants C and C':

$$Pr\{h_2 \to h_1\} \sim C[\lambda\varepsilon]^{\lceil(1-\alpha)s\rceil \wedge \lceil(1-\beta)s\rceil},$$

$$Pr\{h_1 \to h_2\} \sim C'[(1-\lambda)\varepsilon]^{\lceil\alpha s\rceil \wedge \lceil\beta s\rceil}. \tag{5.6}$$

Let $\nu^{\varepsilon,\lambda}(h)$ be the stationary distribution as a function of ε and λ. For a fixed $\varepsilon > 0$, convention h_2 has high probability when λ is close to zero, whereas convention h_1 has high probability when λ is close to one. However, if we think of λ as a *fixed* propensity to play strategy 1, then for all sufficiently small positive ε, the process favors convention h_1, which is the risk-dominant convention. In other words, when the noise is sufficiently small, the existence of an individual bias toward playing one strategy or another is washed out by the social bias toward playing the risk-dominant equilibrium.

This observation generalizes as follows. Let $P^{m,s,\varepsilon}$ be adaptive learning on the n-person game G with finite strategy space $X = \prod X_i$ and population classes $C_1, C_2, \ldots, C_n$. Suppose that whenever an i agent (an agent playing role i) is in the error mode in a given state $h \in X^m$, he chooses an action according to the conditional probability distribution $\lambda_i(x_i \mid h) \in \Delta_i$. The distribution $\lambda_i(\ \mid h)$ can be interpreted as his propensity to play various actions when he is not in a rational mood and the state is h. The stochastically stable states are independent of the conditional distributions $\lambda_i(\cdot \mid h)$ so long as they are time-homogeneous and have full support.

We now consider an alternative model of noise, in which the probability of deviating from best reply depends on the prospective loss in payoff from such a deviation, an approach pioneered by Blume (1993, 1995a, 1995b). Let G be a symmetric 2×2 game with payoff matrix

	1	2
1	a, a	c, d
2	d, c	b, b

$a > d, b > c.$

Consider the playing the field model discussed in the preceding chapter, where there is a single population consisting of m individuals. A *state* is an integer k satisfying $0 \leq k \leq m$, that specifies the number of individuals currently playing strategy 1. In state k, let $\pi_i(k)$ be the payoff from playing action i against the field, where we assume for analytical convenience that a player plays himself. Thus,

$$\begin{aligned} \pi_1(k) &= ak/m + c(1 - k/m) \\ \pi_2(k) &= dk/m + b(1 - k/m). \end{aligned} \tag{5.7}$$

Let $(p_k, 1 - p_k)$ be the probabilities with which a representative agent plays strategies 1 and 2 in period $t+1$, given that k is the state in period t. We shall suppose that the *propensity* to play an action is exponentially related to its expected payoff; that is, for some positive number β,

$$p_k = e^{\beta\pi_1(k)}/[e^{\beta\pi_1(k)} + e^{\beta\pi_2(k)}], \quad \beta > 0. \tag{5.8}$$

We shall call this the *log linear response model* with parameter β, or simply the *β-response model*. To understand the difference between this model and adaptive play, in which errors have uniform probability ε, suppose that in state k, strategy 2 is the best reply, that is, $\pi_1(k) < \pi_2(k)$. Let $\Delta(k) = \pi_2(k) - \pi_1(k) > 0$ and let $\varepsilon = e^{-\beta}$. When β is large, (5.8) implies that

$$p_k = e^{-\beta\Delta(k)}/(1 + e^{-\beta\Delta(k)}) \cong e^{-\beta\Delta(k)} = \varepsilon^{\Delta(k)}. \tag{5.9}$$

In other words, the β-response rule is a perturbed best-reply process in which the probability of *not* choosing the best reply is approximately

$\varepsilon^{\Delta(k)}$, where $\Delta(k)$ is the loss in payoff from playing the second-best strategy.

We claim that in symmetric 2×2 games, this leads to the same long-run outcome as in the uniform error model. Let $P^{m,\beta}$ denote the Markov process on the state space $\{0, 1, 2, \ldots, m\}$ generated by the β-response model. This process is clearly irreducible, so it has a unique stationary distribution $\nu^{m,\beta}$. If action 1 is the best reply in state k, the probability is high that the agent will choose action 1 when β is large. In other words, the β-response rule approaches the best-reply rule as $\beta \to \infty$. In keeping with our earlier definition, we say that a state of the system is stochastically stable if $\lim_{\beta\to\infty} \nu^{m,\beta}(k) > 0$.

In each period, the process either stays in its current state or moves to a neighboring state. Hence each state is associated with a unique rooted tree, and one can estimate the stationary distribution as a function of β and the population size m, using the method described in Section 4.5. This analysis shows that when β and m are large, $\nu^{m,\beta}$ puts almost all the probability on a neighborhood of the risk-dominant convention. That is, for any $p < 1$ and any $\delta > 0$, the probability is at least $1 - \delta$ that at least the proportion p of the population is playing the risk-dominant equilibrium whenever m and β are sufficiently large. Thus the two models of noise yield similar asymptotic results in symmetric 2×2 games. This equivalence does not extend to general games, however; indeed, it is an open problem to characterize the stochastically stable outcomes of the β-response model in more general situations, and to contrast the model's predictions with those of the uniform error model.

5.4 Unbounded Memory

Let us now consider a variation of the learning process in which agents best-respond to samples of past actions and they occasionally make errors, but memory is unbounded (that is, they sample from all past actions). This leads to a nonergodic process whose long-run behavior is qualitatively different from the processes we have considered thus far. For simplicity we shall consider only 2×2 games; the analogous dynamics in general games are more complex.

Let G be a 2×2 matrix game that is nondegenerate. Consider adaptive learning with sample size $s \geq 1$, error rate $\varepsilon > 0$, and infinite memory. Then the state space is the same as in fictitious play, but the process is stochastic instead of deterministic. In particular, there is variability in the players' responses, depending on the samples they happen to draw and whether the players "tremble." Because the sample size is fixed, each player responds to a smaller and smaller fraction of the history as time runs on.

As in the analysis of fictitious play (see Section 2.3), we shall study the motion of the process when it is projected into the space of population proportions. Each player has two actions, which we shall call A and B. Let the random variable $\hat{p}_1^t$ be the proportion of row players who have played A up through time t, and let $1 - \hat{p}_1^t$ be the proportion who have played B. Similarly, let $\hat{p}_2^t$ be the proportion of column players who have played A up through time t, and $1 - \hat{p}_2^t$ the proportion who have played B. Thus the state is $\hat{p}^t = [(\hat{p}_1^t, 1 - \hat{p}_1^t), (\hat{p}_2^t, 1 - \hat{p}_2^t)] \in \Delta$ at the end of period t.

At the beginning of period $t+1$, each player draws a random sample from the past actions of the other player, all samples being equally likely. (As usual, the identities of the players change over time.) Each player chooses a best reply to his or her sample frequency distribution with probability $1-\varepsilon$, and each chooses an action at random with probability ε. For simplicity, assume that ties in best reply are resolved in favor of A. Define the random vector $\hat{B}^t(\hat{p}^t) = [\hat{B}_1^t(\hat{p}^t), \hat{B}_2^t(\hat{p}^t)] \in \Delta$ as follows:

$$\begin{aligned}\hat{B}_i^t(\hat{p}^t) &= (1,0) \text{ if } i \text{ chooses action A in period } t+1,\\ \hat{B}_i^t(\hat{p}^t) &= (0,1) \text{ if } i \text{ chooses B, } i = 1, 2.\end{aligned}$$

Note that the realization of $\hat{B}_i^t(\cdot)$ depends on the samples that the players happen to draw at time t. Let t_0 be the time at which the process begins, and let p^{t_0} be the initial state. Then for all $t \geq t_0$,

$$\hat{p}^{t+1} = \frac{t\hat{p}^t + \hat{B}^t(\hat{p}^t)}{t+1}. \tag{5.10}$$

This stochastic process can be represented as an urn scheme.[1] Imagine two urns of infinite capacity, where U_1 is the row player's urn and U_2 is the column player's urn (see Figure 5.3). At the end of period t, each urn contains t balls, which are of two different colors. White balls correspond to plays of action A, and black balls correspond to plays of action B. At the beginning of period $t+1$, with probability $1-\varepsilon$, the row player reaches into the column player's urn and pulls out s balls at random. She adds one new ball to her own urn, with the color depending on the

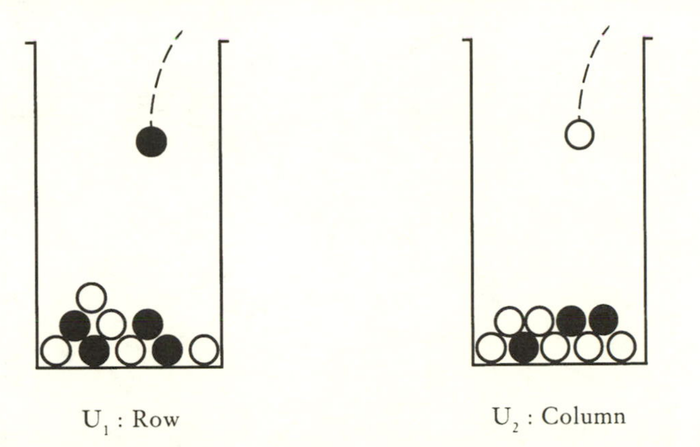

Figure 5.3. Past actions by row and column players represented as colored balls accumulating in two urns.

realization of the random variable $\hat{B}_1^t$, and then she returns the sampled balls to urn U_2. With probability ε, however, she does *not* sample and simply adds one new ball to column's urn, the ball being black or white with equal probability. The column player follows a similar procedure.

We shall analyze this process under the assumption that the game is nondegenerate and has a unique mixed equilibrium $(p_1^*, 1 - p_1^*)$ for the row player and $(p_2^*, 1 - p_2^*)$ for the column player. It is useful to distinguish two cases: (i) G has a unique Nash equilibrium, which is mixed; and (ii) G has three Nash equilibria, two pure and one mixed. We can assume without loss of generality that in the latter case, the two pure equilibria are (A, A) and (B, B). We shall deal with this case first.

The first step is to examine the distribution of the binary random variables $\hat{B}_i^t(\hat{p}^t)$. Define $f_1^{s,\varepsilon}(p_2^t)$ to be the probability that the row player $(i = 1)$ chooses action A in state p^t, and define $f_2^{s,\varepsilon}(p_1^t)$ to be the probability that the column player $(i = 2)$ chooses action A in state p^t. When s is large, the sample proportions are (with high probability) close approximations of the true proportions p_1^t and p_2^t. When $\varepsilon = 0$, and s is large, it follows that the players' actions will, in fact, be best replies to the *true* proportions p_1^t and p_2^t with high probability. When $\varepsilon > 0$ and s is large, each player will choose a nonbest reply with probability approximately equal to $\varepsilon/2$. (When choosing an action randomly, a player may choose the best reply by chance, so the probability of not choosing the best reply is about $\varepsilon/2$.) This means that when s is large, $f_1^{s,\varepsilon}(p_2)$ and $f_2^{s,\varepsilon}(p_1)$ are

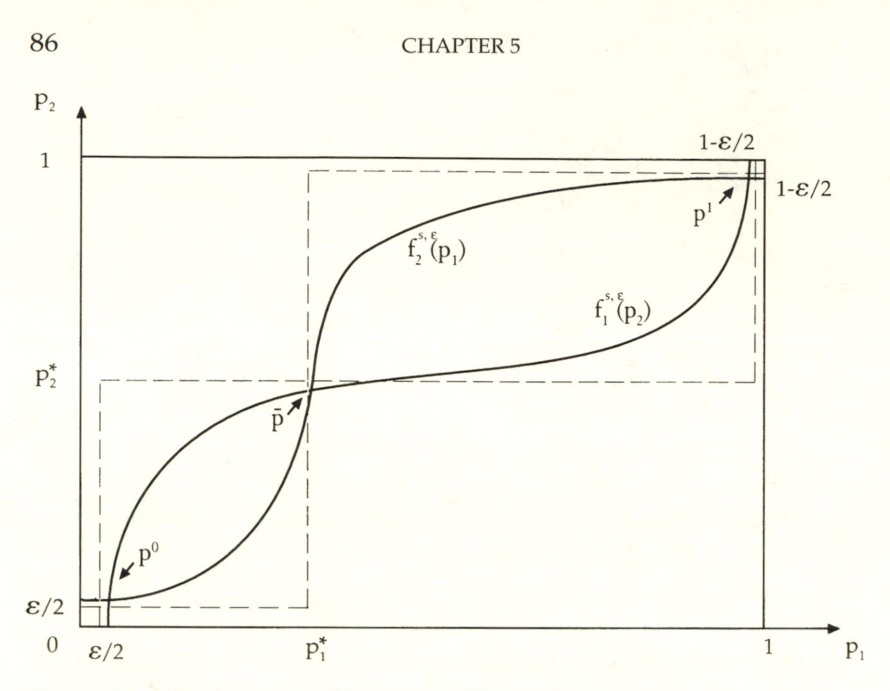

Figure 5.4. The functions $f_1^{s,\varepsilon}(p_2)$ and $f_2^{s,\varepsilon}(p_1)$ when the game has two pure equilibria and one mixed equilibrium.

close to the *perturbed best-reply functions*

$$\phi_1^{\varepsilon}(p_2) = \begin{cases} \varepsilon/2 & \text{if } 0 \le p_2 < p_2^* \\ 1 - \varepsilon/2 & \text{if } p_2^* \le p_2 \le 1 \end{cases}$$

$$\phi_2^{\varepsilon}(p_1) = \begin{cases} \varepsilon/2 & \text{if } 0 \le p_1 < p_1^* \\ 1 - \varepsilon/2 & \text{if } p_1^* \le p_1 \le 1. \end{cases}$$

These step functions are illustrated by the dashed lines in Figure 5.4.

Define

$$f^{s,\varepsilon}(p) = (f_1^{s,\varepsilon}(p_2), 1 - f_1^{s,\varepsilon}(p_2), f_2^{s,\varepsilon}(p_1), 1 - f_2^{s,\varepsilon}(p_1)). \tag{5.11}$$

The *expected* incremental motion of the process is given by the discrete-time dynamical equation

$$\Delta p^{t+1} = \frac{f^{s,\varepsilon}(p^t) - p^t}{t+1}. \tag{5.12}$$

We might suspect that, when t is large, the process behaves like the continuous-time dynamical system

$$\dot{p} = f^{s,\varepsilon}(p) - p. \tag{5.13}$$

This suspicion turns out to be correct. The rest points of the process (5.13) are the points p such that $f^{s,\varepsilon}(p) = p$, which correspond to the points where the graphs of $f_1^{s,\varepsilon}(p_2)$ and $f_2^{s,\varepsilon}(p_1)$ cross. Note that there are three such points, one near each of the Nash equilibria of the game. Using stochastic approximation theory, it can be shown that if the discrete-time stochastic process (5.10) converges, it must converge to one of the rest points of the continuous-time dynamic (5.13). Furthermore, with probability one, it does not converge to the interior rest point that lies near the mixed Nash equilibrium (Kaniovski and Young, 1995).

This result has the following intuitive explanation. The interior Nash equilibrium, which corresponds to the point (p_1^*, p_2^*) in Figure 5.4, is not asymptotically stable in the ordinary fictitious play dynamic, because in the regions to the northeast and southwest, the expected motion is toward one of the corners. If the middle crossing point $(\bar{p}_1, \bar{p}_2)$ is sufficiently close to (p_1^*, p_2^*), the stochastic process (5.10) has probability zero of actually converging to $(\bar{p}_1, \bar{p}_2)$, because the small perturbations due to errors and sampling keep pushing the process into neighboring regions where the motion is toward the corners.

Now consider what happens when the process is close to one of the pure equilibria. Here, sampling variability is small, and the process is mostly buffeted by random errors. These errors can eventually accumulate and push the process out of a neighborhood of one pure equilibrium and into some other part of the space, possibly to a neighborhood of the other pure equilibrium. In fact, it can be shown that no matter how close the process is to a given pure equilibrium at any finite time t, there is a positive probability that it will eventually move to a neighborhood of the other pure equilibrium and never return. However, the probability of these large deviations decreases as the process gets closer to one or the other of the pure equilibria; in fact, from any initial state, the process converges almost surely to one of the fixed points of $f^{s,\varepsilon}(p)$ that lie close to the pure equilibria. Since the probability of converging to each of these points depends on the initial state, the process is nonergodic.

In the case of a 2×2 game with a unique Nash equilibrium, which is strictly mixed, the curves $f_i^{s,\varepsilon}$ have precisely one crossing point $\bar{p} = (\bar{p}_1, \bar{p}_2)$. In this case, the process converges with probability one to $\bar{p}$

from any initial state. We summarize these observations in the following theorem:[2]

THEOREM 5.1. *Let G be a* 2×2 *game and let* $P^{\infty,s,\varepsilon}$ *be adaptive learning with infinite memory, sample size s, and error rate* ε.

(i) *If G has three equilibria, two pure and one mixed, the process converges with probability one to a point that is arbitrarily close to one of the pure equilibria provided s is sufficiently large and* ε *is sufficiently small.*
(ii) *If G has a unique equilibrium, which is mixed, the process converges with probability one to a point that is arbitrarily close to it provided s is sufficiently large and* ε *is sufficiently small.*

This result also holds when other kinds of perturbations are introduced. For example, suppose that the payoffs are random variables that reflect heterogeneity in the agents' utility functions. It can be shown that if the payoffs are independently distributed according to a well-behaved distribution function (say, a normal distribution), then theorem 5.1 continues to hold provided that the variance of the distribution is sufficiently small.

The feature that distinguishes this class of learning processes from the ones we analyzed previously is that the past casts a progressively longer shadow on the present. As a result, the process "slows down" over time: each action taken in period t shifts the state by an amount proportional to $1/t$, and the variability around the expected motion (which arises from errors and incomplete sampling) is also on the order of $1/t$. This results in nonergodic behavior.

While this case is interesting from a theoretical standpoint, it is less compelling as a model of behavior. The fact is that people attach more importance to recent events than to those long past; indeed, precedents may become so "dated" that the players ignore them altogether (or information about them is lost). We have modeled this idea by assuming that memory is finite, which is convenient for analytical purposes. An alternative approach would be to posit that people discount the importance of past actions, or that they sample past actions with a probability that declines exponentially with the number of periods that have elapsed since the action occurred. For a discussion of these and other variants of fictitious play, see Fudenberg and Levine (1998).

5.5 Other Learning Models

The variations considered above fall within the general category of best-reply dynamics. When we depart from this class of learning rules, quite different outcomes may obtain. For example, Robson and Vega-Redondo (1996) study a learning process in which players imitate the behavior of others. Specifically, they consider a *symmetric* 2×2 coordination game that is played by a single population consisting of an *even* number of agents. Suppose that the coordination equilibrium (A, A) has strictly higher payoffs than does the coordination equilibrium (B, B), but that the latter is risk dominant. (The work-shirk game is an example.) Imagine that within each time period, the agents play rounds of the game and that they are rematched at random before each round. Within each period, each agent is committed to playing a fixed action (A or B) for all rounds. At the end of a period, each agent observes the actions of the others and their attendant payoffs. In the simplest version of the model, each agent adopts with probability $1 - \varepsilon$ the action whose average payoff was highest in the previous period, and with probability ε he chooses the action whose payoff was lowest. Ties in payoff can be resolved by the toss of a fair coin.

When $\varepsilon = 0$, a state in which everyone plays the same action is absorbing, because there is only one thing to imitate. When $\varepsilon > 0$, the process can tip from one coordination equilibrium to the other. To compute these tipping probabilities, we use the methodology developed in Chapter 3. Suppose for simplicity that only one round is played in each period. (This does not change the analysis in any important way.) Consider the state in which everyone is playing action B (the risk-dominant equilibrium). At the beginning of the next period, the probability is on the order of ε^2 that two players will adopt action A and that these two players will be matched. Everyone will then think that action A yields higher payoffs than action B, so in the next period everyone (except those who make an error) will choose action A. It follows that the resistance to transiting from the all-B state to the all-A state equals 2 independently of the population size.

Suppose, on the other hand, that the process is in the all-A state and that two agents then mutate into B-players. If the mutants are matched with each other, the average payoff to action B is lower than the average payoff to action A, so the process reverts to the all-A state. If the two mutants are matched against A-players, the *average* payoff to action A still exceeds the *average* payoff to action B provided there are enough other agents (all of whom play A against each other). Thus, when the population size is large enough, it takes *more* than two errors to transit from the all-A state to the all-B state.

This analysis highlights the fact that long-run stability depends on the details of the learning process. While the risk-dominant equilibrium is the stable outcome in a variety of models based on best-reply dynamics, this conclusion does not necessarily hold for other kinds of dynamics. The main point of the evolutionary approach, however, is not to argue for one equilibrium or another on a priori grounds. Rather, it is to show how equilibria arise (and are displaced) when individuals adapt as best they can to complex and changing environments.

Chapter 6

LOCAL INTERACTION

SO FAR we have considered learning models in which agents live in a global soup and interact with each other purely at random. In reality, however, people are separated by location, language, culture, occupation, and a thousand other variables that affect the chances of their meeting. In this chapter we show how to incorporate local interaction effects into the analysis. We begin by discussing a variation of adaptive play in which each agent interacts *mainly* with agents who live nearby. Then we consider the case where each agent interacts *exclusively* with a fixed group of neighbors.

Consider a symmetric two-person game G with finite action set X_0 played by a single population of m agents. Let $u(x, x')$ be the utility to the row player of the action pair (x, x'), and let $u(x', x)$ be the utility to the column player. Assume that the players are embedded in some space that allows us to define the social or geographic "distance" between any two individuals. (The details of this distance function are not important here.) Instead of representing states as histories of previous choices, it is more convenient in this context to represent the state as the current (or most recent) choice of action by each individual in the population. Specifically, the *state* at a given time t is a vector $\mathbf{x}^t \in X_0^m$, where $x_i^t \in X_0$ is the current choice of action by player i, $1 \leq i \leq m$.

The process operates very similarly to the playing the field model described in section 4.4. At the beginning of period $t+1$, one agent is drawn at random from the population. He then draws a sample of s agents (without replacement) from the general population, and computes the sample frequency distribution $\hat{p}^t \in \Delta(X_0)$ of actions that they took in period t. The agent chooses a best reply to $\hat{p}^t$ with high probability, and chooses a nonbest reply with low probability. The new element is that the probability of drawing a particular sample depends on where the sampling agent lives in relation to those he is sampling. The intuitive idea is that a person is more likely to hear about actions that occur nearby than those that occur far away. Since the spatial relationship among the agents is fixed, the probability that a given agent draws any particular sample is also fixed.

Consider now a hypothetical one-period transition between two states, say $x \to x'$. For this transition to occur in either the perturbed or unperturbed process, there must be exactly one agent i who makes a choice x_i' (possibly $x_i' = x_i$), while all other agents stay with their previous choices. In the unperturbed process, the feasibility of the transition depends on what samples i can draw with positive probability and whether x_i' is a best reply to one or more of these samples. It follows that the feasibility of a transition depends only on the *support* of i's sampling distribution. Hence the recurrent classes of the unperturbed process depend only on the support of the agents' sampling distributions.

Now consider a transition $x \to x'$ in the perturbed process. The resistance of this transition is determined by the greatest probability that i would choose x_i' among all possible realizations of i's sample frequency distribution $\hat{p}^t$ in state x. This too depends only on the support of i's sampling distribution. Hence the stochastically stable states of such a process depend only on the support of the agents' sampling distributions. Suppose now that agents gain information *mainly* from people who live nearby, but that sometimes they interact with people who live far away. Under this scenario it is reasonable to assume that every possible sample has a positive probability of being drawn by any given agent, that is, the sampling distributions have full support. If this is the case, then by the preceding argument the stochastically stable states will be *independent* of the particular way in which the agents are spatially distributed. In other words, the topology of the interaction structure does not affect the stochastically stable states. Where the topology does matter is in the *rate* at which the process moves between equilibrium regimes. All else being equal, inertia is much lower when people interact in small close-knit groups than when they interact purely at random, as we shall see in section 6.2.

6.1 Games Played on Graphs

We specialize now to the case where each agent interacts *exclusively* with a fixed group of neighbors.[1] To model this situation, imagine that each agent is situated at a vertex of a graph Γ. The set of m vertices (or "nodes") will be denoted by V, and the set of undirected edges by E. Each undirected edge $\{i, j\}$ has a positive weight $w_{ij} = w_{ji}$ that measures its relative "importance." Vertices i and j are *neighbors* if i and j are linked by an edge, that is, if $\{i, j\} \in E$. The set of all neighbors of a given vertex i is denoted by N_i. In what follows we shall always assume that Γ has no isolated vertices, that is, $N_i \neq \emptyset$ for all i. The *state* of the process is a vector $x \in X_0^m$ such that $x_i \in X_0$ is player i's current action for each $i \in V$. Denote the set of states by Ξ. Each individual i interacts only with his neighbors. In particular, he gains information from his neighbors, and he plays the game with his neighbors. Assume that the one-period payoff to i in state x is the weighted sum of the payoffs from playing each of his neighbors:

$$\nu_i(x) = \sum_{j \in N_i} w_{ij} u(x_i, x_j). \tag{6.1}$$

Concretely, we may think of i playing against neighbor j w_{ij} times in a given period. Alternatively, i plays against neighbor j once each period, and w_{ij} measures the "importance" of the i-j interaction.

Equation (6.1) defines the payoff function of an m-person game played each period, called the *spatial game*. State x is a *Nash equilibrium* of the spatial game if for every i, and every $x' \in X_0$,

$$\sum_{j \in N_i} w_{ij} u(x_i, x_j) \geq \sum_{j \in N_i} w_{ij} u(x', x_j). \tag{6.2}$$

Consider the following example: the players are countries, each of which can choose one of two actions: legislate the left- or the right-hand rule of the road. Represent each country by the vertex of a graph, and join two countries by an edge if they share a common border. Let us weight each edge by the amount of cross-border traffic between the two countries. Thus, if a country chooses the left rule, its payoff equals the total weight on all edges linking it to neighbors who also use the same rule. In other words, the payoff depends linearly on the number of vehicles crossing the border that do not have to switch sides of the road.

A specific example is illustrated in Figure 6.1, where for simplicity all borders are assumed to have equal weight (and the weights are omitted). Since everyone has an odd number of neighbors, a state is a pure Nash equilibrium of the spatial game if and only if each country adopts the rule followed by the majority of its neighbors. There are four different pure equilibrium patterns (the ones shown in the figure), and sixteen pure equilibria in all. All but two of them are spatially heterogeneous, in which different conventions coexist side by side.

The spatial model described above is static. We shall now embed it in an adaptive learning process of the type discussed in previous chapters, following the approach of Blume (1993). Consider a spatial

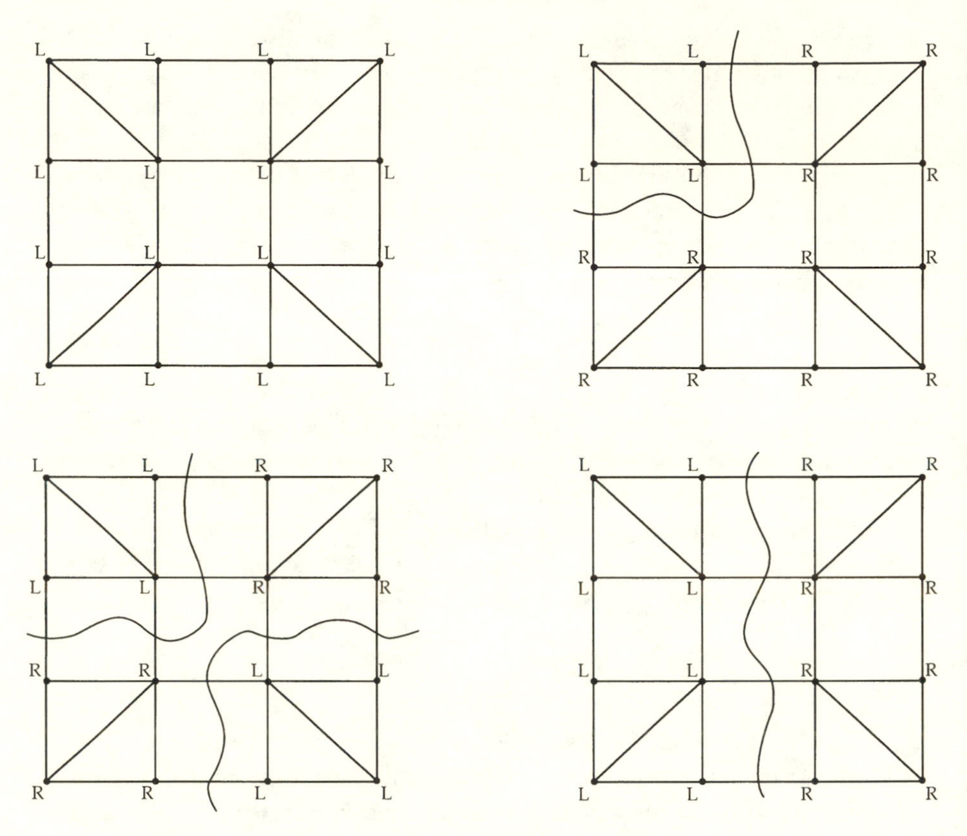

Figure 6.1. Types of equilibria in a spatial game with sixteen vertices.

game defined on the graph Γ, and let x^t be the state at the end of period t. At the beginning of period $t + 1$, one agent (i.e., vertex) i is drawn at random, and i chooses an action $z \in X_0$ according to the probability distribution $p_i^\beta(z \mid x^t)$, where for some $\beta > 0$,

$$p_i^\beta(z \mid x^t) \propto e^{\beta v_i(z, x^t_{-i})}. \tag{6.3}$$

This is a spatial version of the log-linear response rule discussed in Chapter 5. When the coefficient β is large, it amounts to a best-reply process with a small perturbation, where the probability of choosing a nonbest reply declines exponentially with the loss in payoff attendant on choosing it.

Let $P^{\Gamma,\beta}$ be the transition matrix of the associated Markov process. It is clear that $P^{\Gamma,\beta}$ is irreducible, so it has a unique invariant distribution

$\mu^{\Gamma,\beta}(\boldsymbol{x})$. The problem at hand is to estimate the limiting distribution $\lim_{\beta\to\infty}\mu^{\Gamma,\beta} = \mu^{\Gamma,\infty}$. The state or states on which $\mu^{\Gamma,\infty}$ puts positive probability are stochastically stable. Over the long run, they will be observed with much higher probability than will other states.

The computation of the limiting distribution is particularly easy if the underlying game is a potential game. Recall that G is a potential game if there exists some real-valued function $\rho(x, y)$, and a rescaling of the utility functions, such that whenever a player deviates unilaterally, the change in payoff equals the change in potential. For a symmetric two-person game this amounts to saying that there exists a symmetric potential function $\rho(x, y) = \rho(y, x)$ such that for some rescaling of u and for all $x, x', y \in X_0$,

$$u(x, y) - u(x', y) = \rho(x, y) - \rho(x', y).$$

If G is a potential game, then so is the associated spatial game on any weighted graph. To see this, let $\boldsymbol{x}$ be a state of the spatial game, and suppose that agent i deviates, say, by choosing x_i'. Let $\boldsymbol{x}' = (x_i', x_{-i})$. Then

$$\begin{aligned} v_i(\boldsymbol{x}) - v_i(\boldsymbol{x}') &= \sum_{j\in N_i} w_{ij}[u(x_i, x_j) - u(x_i', x_j)] \\ &= \sum_{j\in N_i} w_{ij}[\rho(x_i, x_j) - \rho(x_i', x_j)] \\ &= \sum_{\{h,k\}\in E} w_{hk}\rho(x_h, x_k) - \sum_{\{h,k\}\in E} w_{hk}\rho(x_h', x_k'). \end{aligned}$$

It follows that

$$\rho^*(\boldsymbol{x}) = \sum_{\{h,k\}\in E} w_{hk}\rho(x_h, x_k)$$

is a potential function for the spatial game.

THEOREM 6.1. *Let G be a symmetric potential game with potential function ρ, and let Γ be a finite, weighted graph. For every $\beta > 0$, the spatial adaptive process $P^{\Gamma,\beta}$ has the unique stationary distribution*

$$\mu^{\Gamma,\beta}(\boldsymbol{x}) = e^{\beta\rho^*(\boldsymbol{x})} \Big/ \sum_{\boldsymbol{z}\in\Xi} e^{\beta\rho^*(\boldsymbol{z})}, \tag{6.4}$$

and the stochastically stable states of the spatial game are those that maximize $\rho^*(\boldsymbol{x})$.

The stationary distribution $\mu^{\Gamma,\beta}(\boldsymbol{x})$ is an instance of a Gibbs distribution, a concept that plays an important role in statistical mechanics (Liggett, 1985).

Since the proof of theorem 6.1 is quite short, we shall give it here. For notational simplicity, write μ and P instead of $\mu^{\Gamma,\beta}$ and $P^{\Gamma,\beta}$. The *detailed balance condition* states:

$$\mu(\boldsymbol{x})P_{\boldsymbol{xy}} = \mu(\boldsymbol{y})P_{\mathbf{yx}} \text{ for all } \boldsymbol{x}, \boldsymbol{y} \in \Xi. \tag{6.5}$$

We claim this condition holds when $\mu = \mu^{\Gamma,\beta}$ is defined as in (6.4). Note first that $P_{\boldsymbol{xy}} = P_{\boldsymbol{yx}} = 0$ unless $\boldsymbol{x} = \boldsymbol{y}$ or $\boldsymbol{x}$ and $\boldsymbol{y}$ differ at exactly one site i, that is, $y_i \neq x_i$ and $y_j = x_j$ for all $j \neq i$. Since i has probability $1/m$ of being chosen in any given period, it follows that

$$\begin{aligned}\mu(\boldsymbol{x})P_{\boldsymbol{xy}} &= \left[(1/m)\exp\{\beta\rho^*(\boldsymbol{x})\}/\sum_{\mathbf{z}\in\Xi}\exp\{\beta\rho^*(\boldsymbol{z})\}\right] \\ &\quad\times\left[\exp\left\{\beta\sum_{j\in N_i} w_{ij}u(y_i, x_j)\right\}/\sum_{z_i\in X_0}\exp\left\{\beta\sum_{j\in N_i} w_{ij}u(z_i, x_j)\right\}\right].\end{aligned}$$

Letting

$$\lambda = (1/m)/\left(\sum_{\mathbf{z}\in\boldsymbol{\Xi}}\exp\{\beta\rho^*(\boldsymbol{z})\}\right)\left(\sum_{z_i\in X_0}\exp\left\{\beta\sum_{j\in N_i} w_{ij}u(z_i, x_j)\right\}\right),$$

we have the equivalent expression

$$\begin{aligned}\mu(\boldsymbol{x})P_{\boldsymbol{xy}} &= \lambda\exp\beta\left\{\sum_{\{h,k\}\in E} w_{hk}\rho(x_h, x_k) + \sum_{j\in N_i} w_{ij}u(y_i, x_j)\right\} \\ &= \lambda\exp\beta\left\{\sum_{\{h,k\}\in E} w_{hk}\rho(x_h, x_k) + \sum_{j\in N_i} w_{ij}[u(x_i, x_j) - \rho(x_i, x_j) + \rho(y_i, x_j)]\right\}\end{aligned}$$

$$= \lambda \exp \beta \left\{ \sum_{\{h,k\} \in E} w_{hk} \rho(y_h, y_k) \} + \sum_{j \in N_i} w_{ij} u(x_i, x_j) \right\}$$

$$= \mu(y) P_{yx}.$$

This establishes the detailed balance condition. It follows immediately that μ satisfies the stationarity equation because

$$\sum_{x \in \Xi} \mu(x) P_{xy} = \sum_{x \in \Xi} \mu(y) P_{yx} = \mu(y) \sum_{x \in \Xi} P_{yx} = \mu(y).$$

Since the process is irreducible, it has a unique stationary distribution, which must therefore be μ. This completes the proof of theorem 6.1.

Let us look at two applications of this result. Let G be a 2×2 symmetric game with payoff matrix

	A	B
A	a, a	c, d
B	d, c	b, b

(6.6)

Then G is a potential game, and ρ can be chosen as follows:

$$\begin{array}{ll} \rho(\text{A}, \text{A}) = a - d & \rho(\text{A}, \text{B}) = 0 \\ \rho(\text{B}, \text{A}) = 0 & \rho(\text{B}, \text{B}) = b - c. \end{array} \tag{6.7}$$

Given a state $x \in \Xi$ on a weighted graph Γ, let $w_A(x)$ denote the total weight of all edges such that the agents at both ends choose strategy A. Similarly define $w_B(x)$. Theorem 6.1 says that for every $\beta > 0$ the long-run probability of state x is

$$\mu^{\Gamma,\beta}(x) \propto e^{\beta\{(a-d)w_A(x)+(b-c)w_B(x)\}}. \tag{6.8}$$

Now suppose that G is a coordination game with coordination equilibria (A, A) and (B, B), that is, $a > d$ and $b > c$. Strategy A is strictly risk dominant if and only if $a - d > b - c$, and strategy B is strictly risk dominant if the reverse inequality holds. It follows that, as $\beta \to \infty$, the stationary distribution puts all the support on those states in which everyone is coordinated on the risk-dominant equilibrium. If neither equilibrium is strictly risk-dominant ($a - d = b - c$), a similar argument shows that in every stochastically stable state, there is complete coordination within each connected component of the graph, but that different components may be coordinated on different strategies. We summarize this result as follows:

COROLLARY 6.1. *Let G be a symmetric* 2×2 *coordination game and let* Γ *be a weighted finite graph. In the spatial game, the stochastically stable states of the adaptive process* $P^{\Gamma,\beta}$ *are those in which every connected component is coordinated on a risk-dominant equilibrium.*

The message of corollary 6.1 is that different conventions can persist side by side for the short or intermediate run, but complete coordination within each connected component is the most likely configuration over the long run. In Chapter 1, we pointed out that the evolution of driving conventions in Europe followed a pattern that is not unlike the one predicted by the model. One country (France) changed its convention because of an idiosyncratic shock. Then the new convention spread to neighboring countries, either by voluntary adoption or by force, until by the mid-twentieth century, all countries in continental Europe had adopted the convention of driving on the right. Thus a heterogeneous pattern was eventually displaced by a homogeneous one, though it took over 150 years. Note also the significance of connectedness: the last hold-out in this part of the world is Britain, and given its recent connection to the Continent via the Chunnel, it would not be surprising if it, too, switched eventually.

6.2 Interaction Structure and Speed of Adjustment

Learning with a fixed interaction structure is quite rare in practice, though, as the above example shows, it is not impossible. The more typical situation is that people interact *mostly* with their neighbors. In this case, the stochastically stable states are invariant to the interaction probabilities so long as everyone interacts with everyone else with positive probability (as we argued at the outset of this chapter). What does change, however, is the *rate* at which the system moves from one equilibrium regime to another. When individuals interact mainly with small groups of neighbors, shifts of regime can occur exponentially faster than in the case of uniform interaction. In fact, under certain conditions, the expected waiting time to go from one equilibrium regime to another depends *only* on the size of the groups with which people interact. All else being equal, the smaller the size of the neighborhood groups, and the more close-knit they are, the faster the transition time for the whole population. This means that transitions can occur quite rapidly, even when the error rate is small and the population is large. Ellison (1993) was the first to illustrate this point for the case of agents interacting with their neighbors on a circle. Here, we shall extend the analysis to general interaction structures.

Let G be a symmetric 2×2 game with payoff matrix as in (6.6), and let $\mathcal{G}$ be a class of local interaction structures (graphs) that are structurally similar. For example, $\mathcal{G}$ might consist of all polygons, that is, structures where people are arranged around a ring and each interacts with the person on his left and the person on his right. Or $\mathcal{G}$ might consist of all square lattices embedded on a torus, in which case each person has four neighbors.

We shall be interested in estimating the waiting times as a function of the population size (the number of vertices) for a given class of such structures. For ease of exposition we shall assume that all edges have unit weight. We shall also think of the adaptive process as operating in continuous time. Assume that each individual updates his strategy at random times, which are governed by a Poisson arrival process. In an interval of length τ, the expected number of times a person updates is $\lambda\tau$ for some $\lambda > 0$. There is no loss of generality in assuming that $\lambda = 1$, that is, in assuming that, on average, each individual updates once per unit interval of time. We assume that the updates are independent and identically distributed among individuals. When an individual updates, he does so according to the β-response rule defined in (6.3). Define this continuous-time process by $\bar{P}^{\Gamma,\beta}$. Note that the probability is negligible that any two individuals will update at exactly the same time. Thus with probability one the distinct times at which someone updates define a discrete-time Markov chain that has the same transition probabilities as the discrete-time process $P^{\Gamma,\beta}$.

Assume that the game G has a strict risk-dominant equilibrium, say, equilibrium (A, A). Given any graph in $\mathcal{G}$, theorem 6.1 implies that the stochastically stable state is the configuration in which everyone plays action A. How long does it take (in expectation) for the process to reach this state? The answer depends on such factors as the response coefficient β, the structure of the graph, and the number of vertices.

To gain some insight into this problem, consider an individual who is in the process of updating her choice. The probability that she will *not* choose a best reply is on the order of $e^{-\beta\Delta}$, where Δ is the payoff difference between the best reply and the nonbest reply. The size of Δ depends on the number of people with whom she interacts. Let us suppose that no individual interacts with more than k neighbors in any graph in $\mathcal{G}$. In other words, the size of each person's network is bounded. Then Δ is bounded above by some Δ^*, so the probability that an individual chooses B at any given update is bounded below by some $\varepsilon^* > 0$. It follows that when the total population size (the number of vertices m) is large, the probability that everyone will be playing action A at any given time will be arbitrarily small. Hence the expected waiting time until the process reaches this state can be arbitrarily large.

Nevertheless, we may ask how long it takes for the process to reach a state in which a large *proportion* of the population is playing the risk-dominant equilibrium (see Ellison, 1993). Consider a graph $\Gamma \in \mathcal{G}$ with m vertices. For each state x, let $a(x)$ be the number of individuals playing action A, and let $p(x) = a(x)/m$ denote the proportion of individuals playing action A. Given $p \in [0, 1]$, let $W(\Gamma, \beta, p, \mathbf{x}^0)$ be the expected waiting time until at least $1 - p$ of the population is playing action A, conditional on starting in state $\mathbf{x}^0$:

$$W(\Gamma, \beta, p, \mathbf{x}^0) = E[\min\{\tau\colon p(\mathbf{x}^\tau) \geq 1 - p\}]. \tag{6.9}$$

The *p-inertia* of the process is the maximum expected waiting time over all initial states $\mathbf{x}^0$,

$$W(\Gamma, \beta, p) = \max_{\mathbf{x}^0 \in \Xi} W(\Gamma, \beta, p, \mathbf{x}^0). \tag{6.10}$$

This is achieved when everyone begins by playing the non–risk dominant action B.

We are going to show that $W(\Gamma, \beta, p)$ is determined by the local density of the interaction structure, rather than the size of the graph per se. That is, when individuals belong to small, closely knit groups, the waiting time to get close to the stochastically stable state is bounded above independently of the population size.

To establish this result, we need a formal definition of "closely knit." Given any two nonempty subsets S, S' of the vertex set V, let $e(S, S')$ denote the number of edges such that one end is in S and the other is in

S'. The *degree* of vertex i is $d_i = e(\{i\}, V - \{i\})$, that is, the number of edges that meet vertex i. By assumption, $d_i > 0$ for all i (no vertex is isolated). Given a nonempty subset S of vertices and a real number $0 \leq r \leq 1/2$, we say that S is *r–close-knit* if

$$\forall S' \subseteq S, S' \neq \varnothing, \quad e(S', S) / \sum_{i \in S'} d_i \geq r. \tag{6.11}$$

When $S' = \{i\}$ consists of a single vertex in S, the definition says that at least the fraction r of i's interactions are with other members of S. In this case S is said to be *r-cohesive* (Morris, 1997). This is not a sufficient condition for S to be r-close-knit however. Suppose, for example, that each member of S has *exactly* r of his or her interactions with other members of S. Then S is only $(r/2)$-close knit. The reason is that, with $S' = S$ in (6.11), every interaction between members of S is counted once in the numerator but twice in the denominator. Thus r-cohesiveness only implies $r/2$-close-knittedness. It follows in particular that no set can be more than 1/2-close-knit.

Given a positive integer k and $0 \leq r \leq 1/2$, we say that the graph Γ is *(r, k)–close-knit* if every person belongs to some group of size at most k that is r–close-knit. Similarly, a class $\mathcal{G}$ of graphs is *(r, k)–close-knit* if every graph in the class is (r, k)–close-knit.

As an example, consider the class of all polygons. In a polygon, the degree of every vertex is two. Each subset S of k consecutive vertices contains $k-1$ edges, so $e(S, S)/\sum_{i \in S} d_i = (k-1)/2k$. It is easy to check that in fact $e(S', S)/\sum_{i \in S'} d_i \geq (k-1)/2k$ for every nonempty subset S' of S. It follows that every subset of k consecutive vertices is $(1/2 - 1/2k)$–close-knit. Since every vertex is contained in such a set, the class of polygons is $(1/2 - 1/2k, k)$–close-knit. As a second example, consider the class of square grids embedded on the surface of a torus. It can be verified that a subsquare of size $k = h^2$ is $(1/2 - 1/2h)$–close-knit. Hence the class is $(1/2 - 1/2h, h^2)$–close-knit.

The following result (which is proved in the Appendix) shows that the inertia is bounded provided that everyone is involved in at least one small, sufficiently close-knit group.

THEOREM 6.2. *Let G be a symmetric two-person coordination game with payoff matrix as in (6.6). Assume that equilibrium (A, A) is strictly risk dominant, that is, $a - d > b - c > 0$. Let $r^* = (b-c)/((a-d) + (b-c)) < 1/2$, and let $\mathcal{G}$ be a class of graphs that are (r, k)-close knit for some fixed $r > r^*$ and some* *fixed $k \geq 1$. Given any $p \in (0, 1)$ there exists a β_p such that for each fixed $\beta \geq \beta_p$, the p-inertia of the process $\bar{P}^{\Gamma,\beta}$ is bounded above for all $\Gamma \in \mathcal{G}$, and in particular it is bounded independently of the number of vertices in Γ.*

Earlier we noted that the class of polygons is $(1/2 - 1/2k, k)$–close-knit and that the class of square grids is $(1/2 - 1/2h, h^2)$–close-knit. Given a 2×2 game with a strictly risk-dominant equilibrium, we have $r^* < 1/2$, so we can certainly find integers k and h such that $r^* < 1/2 - 1/2k$ and $r^* < 1/2 - 1/2h$. Theorem 6.2 therefore implies that for any p strictly between zero and one, the p-inertia of graphs in these classes is bounded above independently of the population size.

The intuition behind the theorem can be explained as follows. Consider an (r, k)-close-knit graph Γ. Each individual is contained in an r-close-knit group S of size k or less. The probability that such an individual chooses a nonbest reply in a unit interval of time is bounded below by a positive number that depends on β, k, r, and the payoff matrix, but does not depend on the particular graph Γ. Since these parameters are fixed, the expected waiting time is bounded until the first time that all k members of S play A simultaneously. Once this happens, the close-knittedness of S assures that at every subsequent time, everyone in S plays action A with high probability irrespective of what the players outside of S are doing (assuming that β is sufficiently large). Since every individual is in such a group S, and the process is running simultaneously for all individuals, the waiting time is bounded until the first time that a high *proportion* of all individuals are playing action A.

Note that this argument does not depend on the idea that actions spread by diffusion (which they may indeed do). For example, if a local group switches to action A, it becomes more likely that nearby individuals will also adopt action A, which makes it more likely that *their* neighbors will adopt action A, and so forth. Clearly such a process further reduces the waiting time, but it also requires a fair amount of connectivity in the interaction structure.[2] Theorem 6.2 assumes nothing about connectivity, however. In fact, it applies equally well to graphs that consist of many distinct connected components, each of size k. The driving force behind the result is local reinforcement: if people interact mainly within a small group, any switch by the group to the risk-dominant equilibrium takes a long time to undo, and before that happens, most of the other groups will have switched to the risk-dominant equilibrium, also.

Chapter 7

EQUILIBRIUM AND DISEQUILIBRIUM SELECTION IN GENERAL GAMES

WE TURN our attention in this chapter to evolutionary selection in general finite games. As in previous chapters, our focus will be on the long-run behavior of adaptive learning and the regimes on which it puts nonvanishing probability when the background noise is small. Two features emerge that were not present in the 2×2 case. First, the states that emerge in the long run need not be equilibria of the game; instead, they may represent complex, disequilibrium patterns of behavior. Second, when the process does select an equilibrium (as in a coordination game), it typically selects a unique one that can be characterized analytically. Unlike the 2×2 case, however, these stochastically stable equilibria need not be risk dominant. In fact, some coordination games do not have a risk-dominant equilibrium, but they always have a stochastically stable one.

7.1 COORDINATION GAMES

Let G be an n-person game with finite strategy spaces $X_1, X_2, \ldots, X_n$ and payoff functions u_i: $X \to R$ where $X = \prod X_i$. As usual, $P^{m,s,\varepsilon}$ is adaptive learning with memory m, sample size s, and error rate ε operating on G. Recall that a *convention* is a state of form $h_x = (x, x, \ldots, x)$, where x is a strict Nash equilibrium of G. Some games possess no strict Nash equilibria, while others possess many. In a *coordination game*, for example, each player has the same number K of strategies, which can be ordered so that it is a strict Nash equilibrium for everyone to play their kth strategy, $1 \leq k \leq K$. Each coordination equilibrium, therefore, corresponds to a convention. Adaptive learning selects among these equilibria in the sense that the limit distribution puts almost all the probability on one (or more) conventions whenever the sample size is sufficiently large, information is sufficiently incomplete, and the noise is sufficiently small. (This is a consequence of theorem 7.1 below.) Barring ties in the resistance function, the process selects a unique convention.

The correspondence between stochastic stability and risk dominance that we found in 2×2 coordination games does not, however, extend to the general case. Indeed, risk dominance is not even defined for certain coordination games. To see why, consider a two-person coordination game G in which the strategies are labeled 1 to K. Let the payoffs from the strategy pair (i, j) be a_{ij} for the row player and b_{ij} for the column player. We say that coordination equilibrium (i, i) *strictly risk-dominates* (j, j) in G, written $(i, i)R(j, j)$, if every probability mixture over strategies i and j has only i or j as best replies, and (i, i) strictly risk-dominates (j, j) in the 2×2 subgame obtained by restricting G to strategies i and j (Harsanyi and Selten, 1988). Thus

$$(i, i)R(j, j) \text{ implies } (a_{ii} - a_{ji})(b_{ii} - b_{ij}) > (a_{jj} - a_{ij})(b_{jj} - b_{ji}). \tag{7.1}$$

An equilibrium is *strictly risk dominant* if it strictly risk-dominates all other coordination equilibria in G. Unfortunately, there may be no such equilibrium, as the following example shows:

	1	2	3
1	2, 6	0, 1	1, 0
2	1, 0	3, 4	0, 1
3	0, 1	1, 0	4, 3

Example 7.1. Payoff structure with no risk-dominant equilibrium.

In this example, it is easy to check that (3, 3) R (2, 2), (2, 2) R (1, 1), and (1, 1) R (3, 3), hence none of the equilibria is strictly risk dominant.

Even when a risk-dominant equilibrium exists, it may differ from the stochastically stable equilibrium. Consider the following example:

	1	2	3
1	5, 60	0, 0	0, 0
2	0, 0	7, 40	0, 0
3	0, 0	0, 0	100, 1

Example 7.2. Payoff structure with the risk-dominant equilibrium different from the stochastically stable equilibrium.

Since the off-diagonal payoffs are uniformly zero, the risk-dominant equilibrium is the one that maximizes the product of the payoffs to the row and column players, which in this case is equilibrium (1, 1).[1]

To find the stochastically stable equilibrium, construct a directed graph with three vertices, one for each of the coordination equilibria (see Figure 7.1). Let j denote the vertex corresponding to the equilibrium (j, j), $j = 1, 2, 3$. The corresponding state in the adaptive process is the convention in which everyone has played j for m periods in succession. Denote this state by h_j. For the process to tip into the basin of attraction of convention h_k requires that a certain number of row or column players choose action k by mistake. Ignoring integer problems for the moment, we need to find the minimum proportion p^* such that if p^* of the row players choose k and $1 - q^*$ choose j, then k is a best reply for the column players. Evidently p^* is the solution of $p^* b_k = (1 - p^*) b_j$, that is, $p^* = b_j/(b_j + b_k)$. Similarly, let q^* be the minimum proportion such that if q^* of the column players choose k and $1 - q^*$ choose j, then k is a best reply for the row players. Clearly, $q^* = a_j/(a_j + a_k)$. Letting s be the sample size, it follows that the resistance to the transition $h_j \to h_k$ is

$$r^s_{jk} = \lceil s r_{jk} \rceil, \text{ where } r_{jk} = a_j/(a_j + a_k) \wedge b_j/(b_j + b_k). \tag{7.2}$$

We shall refer to the numbers r_{jk} as *reduced resistances*. The reduced resistances for example 7.2 are illustrated in Figure 7.1. Using the method of rooted trees, we can easily deduce that convention h_2 has minimum stochastic potential. It follows that when s is sufficiently large, and s/m and ε are sufficiently small, the convention "all play 2" will be observed with arbitrarily high probability in the long run. In particular, it will be observed with much higher probability than will the risk-dominant convention "all play 1."

We remark that formula (7.2) for the resistances holds only for coordination games whose off-diagonal payoffs are zero. In other coordination games the computation of the resistances can be more complex. In particular, the resistance to going from the convention "all play i" to the convention "all play j" may involve intermediate strategies different from i and j. This is the case in example 7.1, where the least resistance path from the convention "all play 1" to the convention "all play 2" involves choosing strategy 3 a certain number of times by mistake (as the reader may verify). In general n-person games, the computation of the resistances becomes even more complex; nevertheless, a fair amount

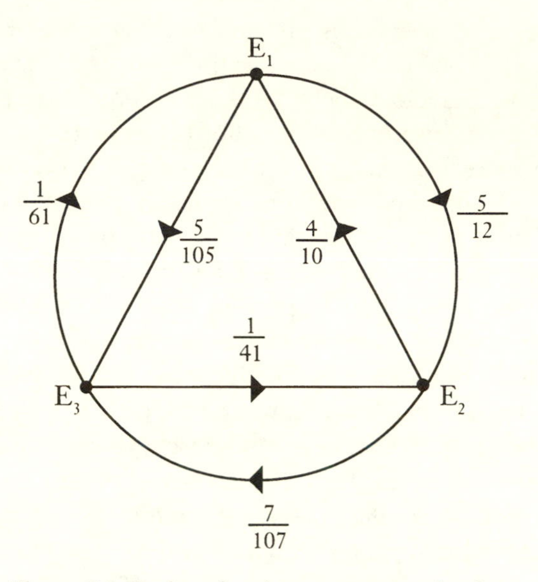

Figure 7.1. Reduced resistances among the three equilibria in example 7.2.

can be said about the stochastically stable states without computing the resistances explicitly, as we show in the next section.

7.2 Weakly Acyclic Games

Given an n-person game G with finite strategy space $X = \prod X_i$, associate each strategy-tuple $x \in X$ with the vertex of a graph. Draw a directed edge from vertex x to vertex x' if and only if there exists exactly one agent i such that $x'_i \neq x_i$, where x'_i is a best reply by i to $x_{-i} = x'_{-i}$. This is called the *best-reply graph* of G (Young, 1993a). A *best-reply path* is a sequence of form $x^1, x^2, \ldots, x^k$ such that each pair (x^j, x^{j+1}) corresponds to an edge in the best-reply graph. A *sink* is a vertex with no outgoing edges. Clearly, x is a sink if and only if it is a strict Nash equilibrium in pure strategies.

Figure 7.2 shows a best-reply graph with two sinks—(C, A) and (B, B). The *basin* of a sink is the set of all vertices from which there exists a directed path ending at that sink. Note that a vertex may be in several basins simultaneously. For example, (B, A) is in the basin of (C, A) and also in the basin of (B, B).

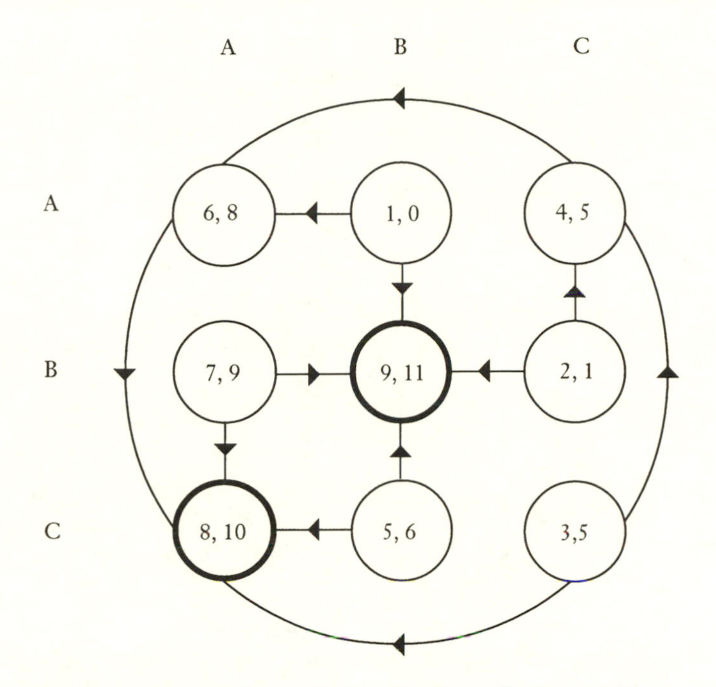

Figure 7.2. Best-reply graph for a 3×3 acyclic game with two sinks.

A game is *acyclic* if its best-reply graph contains no directed cycles. It is *weakly acyclic* if every vertex lies in the basin of at least one sink. It is easily seen that the game shown in Figure 7.2 is acyclic. If we change the payoff pair (8, 10) to (8, 2), and also change (5, 6) to (5, 5), then the graph contains only one sink, (B, B), and the game is weakly acyclic but not acyclic (see Figure 7.3).

Every coordination game is acyclic, because each edge in the best-reply graph points toward a coordination equilibrium. Hence there can be no cycles. Another important class of acyclic games are potential games in which there are no payoff ties. Recall that a *potential game* is one for which there exists a potential function $\rho\colon X \to R$ such that, under an appropriate rescaling of the utility functions, the following holds

$$\forall x \in X, \forall i, \forall x'_i \in X_i, u_i(x'_i, x_{-i}) - u_i(x_i, x_{-i}) = \rho(x'_i, x_{-i}) - \rho(x_i, x_{-i}).$$

This concept can be generalized as follows. Suppose that we require

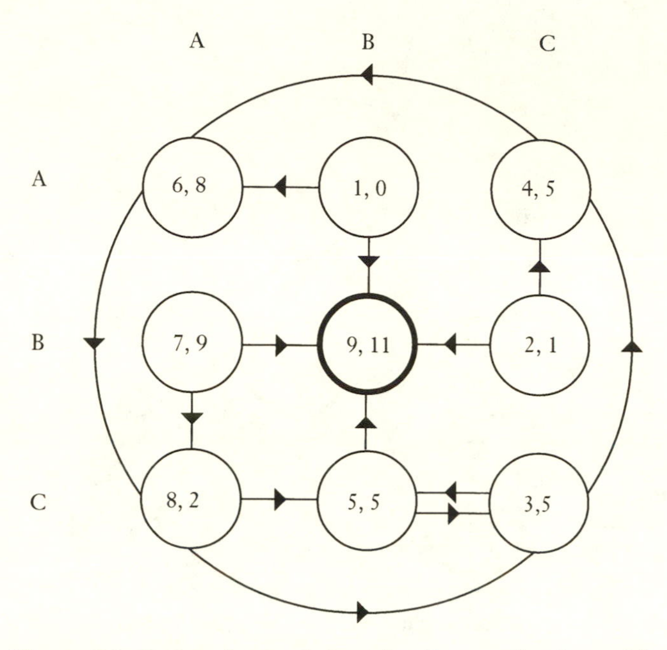

Figure 7.3. Best-reply graph for a 3×3 game that is weakly acyclic but not acyclic.

only that

$$\forall x \in X, \forall i, \forall x_i' \in X_i,$$

$$\text{sign}\{u_i(x_i', x_{-i}) - u_i(x_i, x_{-i})\} = \text{sign}\{\rho(x_i', x_{-i}) - \rho(x_i, x_{-i})\}. \quad (7.3)$$

Then G is an *ordinal potential game* and ρ is an *ordinal potential function* (Monderer and Shapley, 1996b).

We claim that if G is an ordinal potential game in which no player is indifferent between distinct strategies, then G is acyclic. To see this, start at any vertex x^1. If x^1 is a sink (a strict Nash equilibrium), we are done. If x^1 is not a sink, there exists a player i and a strategy $x_i^2 \neq x_i^1$ such that x_i^2 is the unique best reply by i to x_{-i}^1. Let $x^2 = (x_i^2, x_{-i}^1)$. If x^2 is not a sink, there exists some player j whose unique best reply to x_{-j}^2 is $x_j^3 \neq x_j^2$, and so forth. In this manner we construct a best-reply path along which ρ is strictly increasing. Hence it cannot cycle, and must ultimately end at a sink.

THEOREM 7.1. *Let G be a weakly acyclic n-person game, and let $P^{m,s,\varepsilon}$ be adaptive play. If s/m is sufficiently small, the unperturbed process $P^{m,s,0}$*

converges with probability one to a convention from any initial state. If, in addition, ε is sufficiently small, the perturbed process puts arbitrarily high probability on the convention(s) that minimize stochastic potential.

When the sample size s is large, the process will discriminate quite sharply between alternative equilibria; indeed, typically there will exist a unique coordination equilibrium that is stochastically stable for all sufficiently large s.

7.3 CURB SETS

Some games—such as the fashion game—have a structure that naturally yields cyclic behavior; in such cases, we should not expect adaptive learning to settle into an equilibrium. Nevertheless, much can be said about the (disequilibrium) regimes that adaptive processes select in this type of situation. Indeed, theorem 3.1 tells us that certain recurrent classes will be observed with high probability when the perturbations to the learning process are small. What properties do these recurrent classes have?

To investigate this question, let us first consider the following game:

	A	B	C	D
A	3, 3	4, 1	1, 4	−1, −1
B	1, 4	3, 3	4, 1	−1, −1
C	4, 1	1, 4	3, 3	−1, −1
D	−1, −1	−1, −1	−1, −1	0, 0

This game has a single sink (D, D) and a single best-reply cycle: CA → CB → AB → AC → BC → BA → CA. Once the process has entered this cycle, there is no exit. It is straightforward to show that these two regimes—the sink and the cycle—are the only two recurrent classes of adaptive play when there are no perturbations and s is sufficiently large and s/m is sufficiently small.

Which of these two regimes is more probable when small perturbations are introduced? The answer is found by counting the number of errors needed to tip the process out of one regime and into the basin of attraction of the other. Suppose first that the process is in the convention where (D, D) is played repeatedly. To move to the basin of attraction of

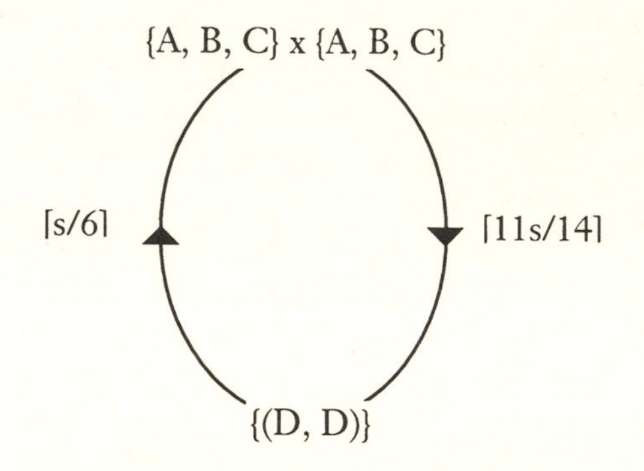

Figure 7.4. A game with a stochastically unstable equilibrium and a stochastically stable disequilibrium.

the cycle requires that either the row or the column player choose some other strategy—say, C—often enough so that D is not the only best reply. If the column player chooses strategy C for $\lceil s/6 \rceil$ periods in succession, then with positive probability the row player's best reply will be B. Assuming that s is small enough relative to m, the process will move into the cycle involving A, B, C. Conversely, to move from the cycle back to (D, D), some player must choose D often enough that it is a best reply. The "weakest" point in the cycle arises when A, B, and C occur with the same frequency in the row (or column) player's sample and D occurs at least eleven times as often as A, B, or C. (The occurrences of D represent mistakes, of course.) For large enough s (and small enough s/m), it is therefore easier to move from the sink to the cycle than the other way around. Hence the cycle is stochastically stable.

This approach generalizes as follows. A *product set* of strategies is a set of form $Y = \prod Y_i$, where each Y_i is a nonempty subset of X_i. Let ΔY_i denote the set of probability distributions over Y_i, and let $\prod(\Delta Y_i)$ denote the product set of such distributions. Let $BR_i(Y_{-i})$ denote the set of strategies in X_i that are best replies by i to some distribution $p_{-i} \in \prod_{j \neq i}(\Delta Y_j)$, that is,

$$BR_i(Y_{-i}) = \{x_i \in X_i \colon \exists p_{-i} \in \prod_{j \neq i}(\Delta Y_j) \text{ s.t. } \sum_{x_{-i}} u_i(x_i, x_{-i})p_{-i}(x_{-i}) \geq \sum_{x_{-i}} u_i(x_i', x_{-i})p_{-i}(x_{-i}) \text{ for all } x_i' \in X_i\}.$$

Let

$$BR(Y) = [BR_1(Y_{-1}) \times \cdots \times BR_n(Y_{-n})]. \tag{7.4}$$

Observe that BR maps product sets to product sets. The product set $Y = \prod Y_j$ is a *curb set* (Closed Under Rational Behavior) if $BR(Y) \subseteq Y$ (Basu and Weibull, 1991). It is a *minimal curb set* if it is curb and contains no smaller curb set. In this case Y is a fixed point of BR: $BR(Y) = Y$. This concept is a natural generalization of a strict Nash equilibrium, which is a *singleton* minimal curb set. Every minimal curb set is a fixed point of BR, but there may exist fixed points of BR that are not minimal. Indeed, the set of rationalizable strategies corresponds to the unique *maximal* fixed point of BR (Bernheim, 1984; Pearce, 1984). In the example shown in Figure 7.4, BR has two minimal fixed points: {A, B, C} × {A, B, C} and {(D, D)}. It also has one maximal fixed point, namely, {A, B, C, D} × {A, B, C, D}.

Given a finite n-person game G, let $P^{m,s,\varepsilon}$ be adaptive play and let $H = X^m$ be the space of truncated histories. The *span* of a subset $H' \subseteq H$, denoted by $S(H')$, is the product set of all strategies that appear in some history in H'. H' is a *minimal curb configuration* if its span coincides with a minimal curb set. We say that a property holds *generically* for a given class of games if it holds for an open dense subset of that class in $R^{n|X|}$. In particular, given any game G in the class, the property holds for "almost all" slight perturbations of the payoffs in G. The following result is a variant of a theorem of Hurkens (1995).[2]

THEOREM 7.2. *Let G be a generic n-person game on the finite strategy space X, and let $P^{m,s,\varepsilon}$ be adaptive play. If s/m is sufficiently small, the unperturbed process $P^{m,s,0}$ converges with probability one to a minimal curb configuration. If, in addition, ε is sufficiently small, the perturbed process puts arbitrarily high probability on the minimal curb configuration(s) that minimize stochastic potential.*

The requirement that G be generic is necessary, as we show by example in the Appendix. We also remark that when the sample size is sufficiently large, there will typically be a unique minimal curb configuration that minimizes stochastic potential, so the selection criterion is sharp.

7.4 Elimination of Strictly Dominated Strategies

A strategy x_i is said to be *strictly dominated* by strategy y_i if y_i yields player i a strictly higher payoff than x_i for any choice of strategies by the other players. It is reasonable to suppose that a rational player will never use a strictly dominated strategy. For each i, let X_i' be i's strategy space after deleting all strictly dominated strategies. This leads to the reduced strategy space $X' = \prod X_i'$. Suppose that every player knows the other players' utility functions, and knows that they behave rationally. Then they can foresee that only strategies in X' will be played.

In X' there may exist strictly dominated strategies that were not strictly dominated in X. After eliminating these strategies from X', we obtain a smaller space X'', and so forth. This process is known as the *iterated elimination of strictly dominated strategies*. The nested sequence $X \supset X' \supset X'' \supset \cdots$ has a minimal element X^* in which no strategy is strictly dominated for any player. If the agents know the others' utility functions, know that the others are rational, know that the others know this, and so forth, up to some suitable level of mutual knowledge, then (and only then) can they deduce that strategies outside of X^* will not be played.

Adaptive learning yields the same result without assuming the mutual knowledge of either utility or rationality. To see why, consider the unperturbed process $P^{m,s,0}$ and let $h \in X^m$ be an arbitrary state. In each period, every agent chooses a best reply to a sample of information about the actions of the other agents in previous periods. Since a strictly dominated strategy is not a best reply to any such sample, it will never be chosen. Thus within m periods, the process reaches a state whose span is contained in X'. Continuing this reasoning, we see that the process must eventually reach a state whose span is contained in X^*. Moreover this must happen within $m|X|$ periods, beginning from any initial state. It follows that the span of every recurrent class of $P^{m,s,0}$ is contained in X^*; that is, all iterated strictly dominated strategies have been eliminated. By theorem 3.1, the recurrent classes of $P^{m,s,0}$ are the only candidates for being stochastically stable. Hence adaptive learning eliminates iterated strictly dominated strategies, but without any assumption of mutual knowledge.[3] In other words, the evolutionary process can substitute for high levels of knowledge and deductive powers on the part of individuals. This is one of the central messages of the evolutionary approach.

Chapter 8

BARGAINING

8.1 Focal Points

Imagine two people who get to divide a pot of money provided they can agree on how to divide it. What proportions will they settle on? In laboratory experiments with this structure, subjects usually divide the money fifty-fifty (Nydegger and Owen, 1974). Fifty-fifty is a natural focal point when parties have an equal claim to a good; indeed, it is hard to imagine how they could justify any other division. But in real-world bargaining, people almost never do have an equal claim to the things being divided. They differ in their needs, contributions, abilities, tastes, and a variety of other characteristics that might well justify unequal shares.

Consider, for example, a negotiation between a principal and an agent about how to divide their joint output. A concrete case would be an agreement between a lawyer and a client in a malpractice case about how to split the jury award (an arrangement known as a contingency fee). In such cases, there is no compelling argument for equal division, because the parties contribute different things to the relationship. In the United States, for example, the standard arrangement is for the lawyer to get one-third and the client two-thirds of the jury award. What is striking is that these proportions have the status of a convention: they seem to be virtually independent of the amount of the award, the intrinsic merit of the case, or the amount of effort the lawyer puts into defending it. Similar practices characterize other principal-agent relationships: tipping percentages in restaurants, franchising fees, real-estate commissions, sharecropping contracts, and so forth.

Conventions like these have economic value: by coordinating the parties' expectations, they reduce transaction costs and the risk that the bargaining process may break down. Conversely, where no clear convention exists, one would expect transaction costs to increase. This point has been demonstrated in laboratory experiments. For example, Roth and Murnighan (1982) examined the following situation: Two subjects were given one hundred lottery tickets to divide. Each subject's chance of winning a prize was proportional to the number of lottery tickets that he or she received in the bargain, but the money value of the prizes was different for the two players. Player A's prize was worth $20, whereas player B's prize was worth $5. Thus if the two players divided the tickets in the proportions 20:80, for A and B, respectively, A would have a 20 percent chance of winning $20, whereas B would have an 80 percent chance of winning $5, so the expected money payoff would be the same for both players. Assuming both players know the value of both prizes, 20:80 is an obvious focal point. But there is also a second focal point, namely, to split the tickets equally. (Indeed, there is a third focal point, which is to split the difference between the equal-money and equal-tickets focal points.) Note, however, that when neither player knows the value of the other person's prize, equal division of the tickets is the unique focal point.

The experiment was run under a variety of scenarios in which one or both players knew the value of the other's prize. There were also two variants of each scenario in which the informational situation was either common knowledge or was not common knowledge. (Information is said to be common knowledge if everyone knows it, everyone knows that everyone knows it, everyone knows that everyone knows that everyone knows, and so forth.) The common knowledge situation was created by telling the bargainers that they were receiving exactly the same instructions.

The experimental results lend support to the hypothesis that when there is more than one focal point, the bargainers are less likely to reach agreement than when there is a unique focal point. For example, when both players knew both prizes (so there were two focal points in both the common knowledge and noncommon knowledge situations), the overall failure rate was 22 percent (14 out of 65 trials). When neither player knew both prizes (so there was a unique focal point in both knowledge situations), the overall failure rate was 11 percent (7 out of 63 trials). When data are pooled from a variety of similar experiments, the failure rate for situations with two focal points is 23 percent (53 out of 226), whereas the failure rate for situations with a unique focal point is 7 percent (10 out of 149). While we cannot safely apply tests of significance to these pooled results because the experiments did not fully control for all factors other than the number of focal points, the data are consistent with the hypothesis that a multiplicity of focal points increases the probability that the bargainers will fail to reach agreement (Roth, 1985).

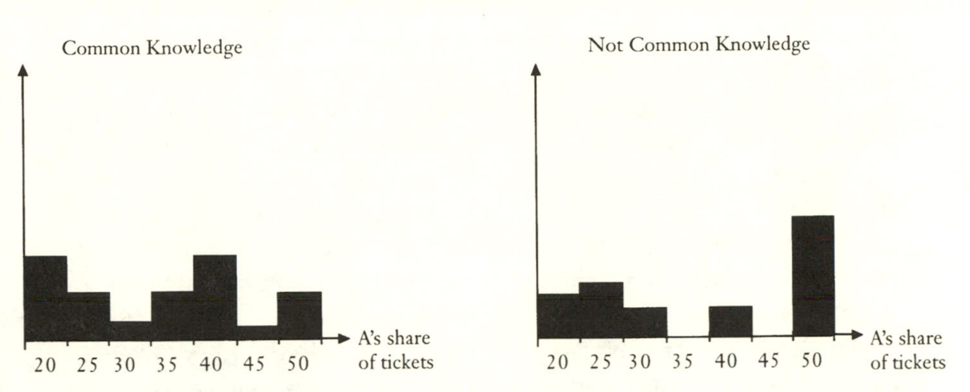

Figure 8.1. Frequency distribution of agreements when both players know both prizes. *Source*: Roth and Murnighan, 1982.

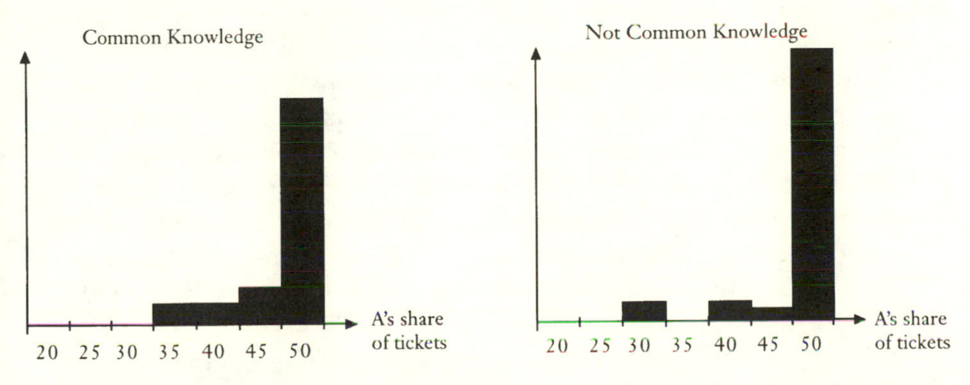

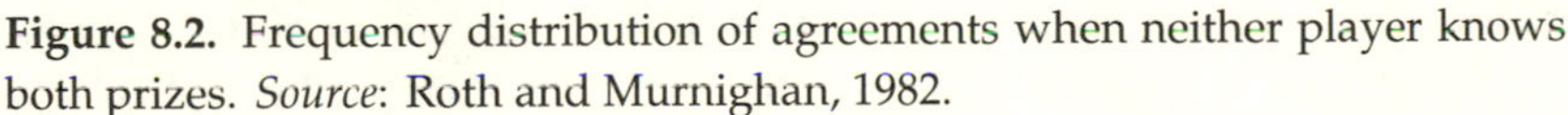

Figure 8.2. Frequency distribution of agreements when neither player knows both prizes. *Source*: Roth and Murnighan, 1982.

There is also a strikingly different *pattern* of agreements when there are two focal points as compared to one focal point. With two focal points, the agreements were spread out between the focal points (see Figure 8.1), whereas with a unique focal point, the agreements were concentrated at the focal point, with some variation around it (see Figure 8.2). Similar results have been obtained in other experiments (see, e.g., Roth and Malouf, 1979; Roth, Malouf, and Murnighan, 1981).

These results lend support to the proposition that a bargaining norm is a form of social capital: it has economic value because it facilitates coordination. But how do bargaining norms become established in the first place? In previous chapters we have argued that conventions arise through the accumulation of precedent: people come to expect a certain division of the pie because other people have agreed to divide the pie in a similar way under similar circumstances. In the case of bargaining, this feedback loop between precedents and expectations can be demonstrated in the laboratory, that is, bargainers can be *conditioned* by precedent to favor one focal point over another through the reinforcement effect of precedent. Roth and Schoumaker (1983) demonstrated this in a series of experiments that paralleled those of Roth and Murnighan (1982). Each pair of subjects was given one hundred lottery tickets to divide. The prize was $40 for A and $10 for B. Since this was common knowledge, there were two focal points: dividing the tickets 50:50 and dividing them 20:80. Each subject played twenty-five games in succession. Unknown to the subjects, however, the first fifteen plays were against a computer that was programmed to play one or the other focal point. After the first fifteen rounds, each subject was paired against a succession of other subjects who had been playing against the same programmed opponent. Furthermore, the solutions that each subject had agreed to in the previous five rounds were published for both to see. The conjecture was that the experience of the subjects in the programmed rounds would create mutual expectations that would tend to lock them into whatever solution they had become accustomed to playing. (A control group of subjects played the full twenty-five rounds without programmed opponents.)

The results confirmed the hypothesis that expectations formed in early rounds of play strongly influence players' subsequent behavior. Almost all of the players who became accustomed to an opponent who insisted on 50:50 in the first fifteen rounds continued to make 50:50 agreements in the remaining ten rounds. This was true even though subjects in role B would have been better off under the alternative 20:80 norm. Similarly, the players in role A who had become accustomed to playing opponents who insisted on the 20:80 division continued to make agreements of this sort even though the alternative 50:50 norm would have been preferable.

These results provide empirical evidence for two general propositions: a bargaining norm has economic value as a coordination device, and the choice of norm can be conditioned through precedent. In the next few sections we shall draw out the implications of these propositions, using the framework developed in previous chapters. Before doing so, however, let us recall what classical bargaining theory has to say about these situations. In the bargaining model originated by Nash

(1950), the outcome of a two-person distributive bargain depends only on the utility functions of the two parties (their attitudes toward risk), and their alternatives if they fail to reach agreement. Let $u(x)$ be the row player's von Neumann–Morgenstern utility function, which we assume is concave and strictly increasing in the row player's share x of the pie. Similarly, let $v(y)$ be the utility function of the column player as a function of that player's share y. Assume that if the bargaining breaks down, their shares are x^0 and y^0, respectively, where $x^0 + y^0 < 1$. Let $u^0 = u(x^0)$ and $v^0 = v(y^0)$. The *Nash bargaining solution* is the unique division of the pie that maximizes the product of the utility gains relative to the disagreement alternatives, that is, it is the unique division $(x^*, 1 - x^*)$ that maximizes $[u(x) - u^0][v(1 - x) - v^0]$ subject to $0 \leq x \leq 1$.

This division arises in the subgame perfect equilibrium of the following noncooperative bargaining game (Stahl, 1972; Rubinstein, 1982). Two players take turns making offers to each other (an offer is a proposed division of the pie). First, player A makes an offer, which player B accepts or rejects. If player B accepts the offer, the game is over. If player B rejects the offer, she gets to make a counteroffer, which player A accepts or rejects, and so forth. After each rejection there is a small probability p that the game will cease with no further offers (the negotiations "break down"). It can be shown that this game has a unique subgame perfect equilibrium, and that the outcome is arbitrarily close to the Nash division when p is sufficiently small.

For the players to actually reach this outcome requires, however, that their utility functions be common knowledge and that the structure of the bargaining have the exact form described above. These assumptions strain credulity: if anything is common knowledge, it is that utility functions are almost never common knowledge. Moreover, there is no reason why the parties need to bargain with each other via alternating offers with a fixed probability p of breakdown. The Nash outcome depends crucially on these assumptions.

The model of norm formation that we propose dispenses entirely with common knowledge, common beliefs, and common priors. We posit instead that people take their cues from what other people have actually done before them. If lawyers usually get one-third of the award as a contingency fee, clients come to expect that lawyers will insist on this much, and lawyers come to expect that their clients will accept this. In short, common expectations emerge endogenously through the accumulation of precedent. It is also reasonable to assume that random perturbations jostle these expectations to some extent. We shall show that when all agents have the same sample size, and all agents within each population have the same utility function, the stochastically stable norm corresponds to the Nash bargaining solution. When the populations are heterogeneous in both sample sizes and utility functions, one obtains a generalization of the Nash solution that differs from the Harsanyi-Selten extension of the Nash solution. This result shows, in particular, how high-rationality solutions from classical game theory can emerge in low-rationality environments through the process of social learning.

8.2 Adaptive Learning in Bargaining

Consider two populations of players—landlords and tenants, lawyers and clients, franchisers and franchisees—who periodically bargain pairwise over their shares of a common pie. We shall refer to these (disjoint) populations as "row players" and "column players." Generically, let x denote the share that a row player gets, and let y denote the share that a column player gets. For the moment we shall assume that each population is homogeneous, that is, that everyone in the same population has the same utility function. Let $u(x)$ denote the row players' utility as a function of the share x, and let $v(y)$ denote the column players' utility as a function of the share y, where $x, y \in [0, 1]$. As usual, we assume that u and v are concave and strictly increasing. For simplicity, we shall also assume that the disagreement shares satisfy $x^0 = y^0 = 0$. This involves no real loss of generality, because we could always say that the pie to be divided is the surplus over and above the disagreement shares. Without loss of generality we can normalize u and v so that $u(0) = v(0) = 0$, and $u(1) = v(1) = 1$.

At the beginning of each period, one row and one column player are drawn at random from their respective populations. They play the *Nash demand game*: the row player demands some number $x \in (0, 1]$, and simultaneously the column player demands some number $y \in (0, 1]$. Note that the demands are strictly positive—it makes no sense to "demand" nothing. The outcomes and payoffs are as follows:

Demands	Outcomes	Payoffs
$x + y \leq 1$	x, y	$u(x), v(y)$
$x + y > 1$	$0, 0$	$0, 0$

To keep the state space finite, we shall discretize the strategies by allowing only demands that are expressible in d decimal places, where d is a fixed positive integer. The *precision* of the demands is $\delta = 10^{-d}$. Let $X_\delta = \{\delta, 2\delta, \ldots, 1\}$ denote the finite space of discretized demands.

The evolutionary process is adaptive play with memory m and error rate ε. It will be convenient to express the sample sizes as fractions of m. (The reason for this will become apparent later on.) Let a be the rational fraction of precedents that the row players sample, and let b be the rational fraction of the precedents that the column players sample, where $0 < a, b \leq 1$. To avoid rounding problems, we shall henceforth assume that m is chosen so that am and bm are both integers (this is purely a matter of mathematical convenience). Let (x^t, y^t) denote the amounts demanded by the row player and column player in period t. At the end of period t, the state is

$$h^t = ((x^{t-m+1}, y^{t-m+1}), \ldots, (x^t, y^t)).$$

At the beginning of period $t+1$, the current row player draws a sample of size am from the y-values in h^t. Simultaneously and independently, the current column player draws a sample of size bm from the x-values in h^t.

Let $\hat{g}^t(y)$ be the frequency distribution of demands y in the row player's sample. Thus $\hat{g}^t$ is a random variable with cumulative distribution function

$$\hat{G}^t(y) = \int_0^y \hat{g}^t(z)\, dz.$$

With probability $1 - \varepsilon$ the row player chooses a best reply given $\hat{G}^t$, that is,

$$x^{t+1} = \operatorname{argmax} u(x)\hat{G}^t(1 - x).$$

With probability ε he chooses a demand at random from X_δ. A similar rule applies to the column players. This yields a Markov process $P^{\varepsilon,\delta,a,b,m}$ on the state space $H = X_\delta^m$.

A *conventional division* or *norm* is a state of form

$$h_x = ((x, 1 - x), (x, 1 - x), \ldots, (x, 1 - x)),$$

where $0 < x < 1$. Within the remembered history, row players have always demanded (and received) x, while column players have always demanded (and received) $1-x$. Thus their actions and expectations are fully coordinated in the absence of errors and other stochastic perturbations. We shall say that a division $(x, 1-x)$ is *stochastically stable* for a given precision δ, if the corresponding norm h_x is stochastically stable for all sufficiently large m such that am and bm are integers.

THEOREM 8.1. *Let G be the discrete Nash demand game with precision δ played adaptively with memory m and sample sizes am and bm, where $0 < a, b \leq 1/2$. As δ becomes small, the stochastically stable division(s) converge to the asymmetric Nash bargaining solution, namely, the unique division that maximizes $(u(x))^a(v(1-x))^b$.*

Note that if both sides have the same utility function ($u = v$) but have different amounts of information ($a \neq b$), the outcome favors the side that is better informed. Note also that the analog of theorem 8.1 holds when players demand discrete shares $x > x^0$ and $y > y^0$: the stochastically stable division comes arbitrarily close to the unique division (x, y) that maximizes $(u(x) - u(x^0))^a(v(y) - v(y^0))^b$ subject to $x > x^0$, $y > y^0$, and $x + y = 1$.

We stress that this result does not depend on agents' calculations about others' utility functions, their information, or the extent of their rationality. None of these things is common knowledge (or even mutual knowledge). The only thing we assume is that agents respond more or less rationally to concrete information about actions taken by their predecessors. Since their responses depend on their attitudes toward risk, the outcome does in fact depend on their utility functions (even though the agents may not be aware of it). The Nash solution is favored over the long run simply because it is the most *stable* given the agents' preferences and the random forces that constantly buffet the process about.

8.3 SKETCH OF THE PROOF OF THEOREM 8.1

We shall outline the proof of theorem 8.1 under two simplifying assumptions:

(i) The functions u and v are continuously differentiable;
(ii) All errors are *local* in the sense that when a player chooses randomly, he makes a demand that is within δ of a best reply to some sample.

(A complete proof without these assumptions is given in Young, 1993b.)

To motivate the argument, consider a concrete example. Let the precision $\delta = .1$, and suppose that the current convention is .3 for the row player and .7 for the column player. Then the state looks like this:

$$\begin{array}{ll} & \overbrace{\qquad\qquad\qquad}^{m \text{ periods}} \\ \text{Row player's previous demands:} & .3\ .3\ .3\ \ldots\ .3 \\ \text{Column player's previous demands:} & .7\ .7\ .7\ \ldots\ .7 \end{array}$$

To tip the process into the basin of attraction of another norm requires a succession of errors. Since by assumption all errors are local, the row player can deviate only to .2 or .4, and the column player can deviate only to .6 or .8. Consider the possibility that the row player demands .4 for several periods. This is too high given the current convention, and the players who demand .4 will almost surely fail to make a bargain, at least at first. However, once enough of these errors accumulate, they will eventually change the expectations of future column players. For example, a column player will respond with .6 if there are i instances of .4 in her sample and $v(.6) \geq (1 - i/bm)v(.7)$. In other words, i mistakes by the row player can push the process into the basin of attraction of the norm (.4, .6) if

$$i \geq \lceil bm(v(.7) - v(.6))/v(.7) \rceil. \tag{8.1}$$

Similarly, the column player can push the process into the basin of attraction of (.4, .6) by demanding .6, which is too little given the current convention. This weakness will eventually be exploited by the row player if there are j instances of .6 in the row player's sample and $(j/am)u(.4) \geq u(.3)$, that is, if

$$j \geq \lceil am\, u(.3)/u(.4) \rceil. \tag{8.2}$$

Each $x \in X_\delta$ such that $\delta \leq x \leq 1 - \delta$ corresponds to a norm that, for brevity, we shall denote by $(x, 1 - x)$. The *resistance* to going from the norm $(x, 1 - x)$ to the norm $(x + \delta, 1 - x - \delta)$ is

$$r(x, x + \delta) = \lceil bm(v(1 - x) - v(1 - x - \delta))/v(1 - x) \rceil \wedge \lceil amu(x)/u(x + \delta) \rceil.$$

When δ is sufficiently small, the first term is the smaller of the two, and it is well approximated by the expression

$$r(x, x + \delta) \approx \lceil \delta bmv'(1 - x)/v(1 - x) \rceil.$$

This has a simple interpretation: if the row player demands δ more than he should, the column player will resist, and the amount of her resistance is her sample size times the relative loss of utility that she suffers by giving up δ of her share. Similarly, if the column player demands δ more than she should, the row player's resistance is his sample size times the relative loss of utility that he suffers by giving up δ. Thus, the resistance of the transition from the norm $(x, 1-x)$ to the norm $(x-\delta, 1-x+\delta)$ is approximately

$$r(x, x-\delta) \approx \lceil \delta amu'(x)/u(x) \rceil.$$

Now construct a graph with one vertex for each discrete norm $(x, 1-x)$, $\delta \leq x \leq 1-\delta$. Denote this vertex by "x." There is a directed edge corresponding to the transition $x \to x-\delta$, and its resistance is approximately $\lceil \delta amu'(x)/u(x) \rceil$. Similarly, the resistance of the directed edge $x \to x+\delta$ is approximately $\lceil \delta bmv'(1-x)/v(1-x) \rceil$. (The resistance to moving from x to a nonneighboring vertex is at least as great as the resistance to moving to one of the neighboring vertices, hence it turns out that we can ignore these transitions.)

Define the function

$$f_\delta(x) = \min\{r(x, x+\delta), r(x, x-\delta)\}.$$

By the preceding argument, the resistance to moving to the right, $r(x, x+\delta)$, is an increasing function of x, whereas the resistance to moving to the left, $r(x, x-\delta)$, is a decreasing function of x. Hence $f_\delta(x)$ is unimodal—first it increases, and then it decreases in x. Let x_δ be a maximum point of $f_\delta(\cdot)$. (There are at most two maxima in the set of discrete demands.) For each vertex x such that $\delta \leq x < x_\delta$, select the directed edge $(x, x+\delta)$, and for each vertex x such that $1-\delta \geq x > x_\delta$, select the directed edge $(x, x-\delta)$. Together, these edges constitute an x_δ-tree, T, because from each vertex other than x_δ there is a unique directed path to x_δ (see Figure 8.3).

We claim that T is the least resistant rooted tree. To prove this, recall that every rooted tree has exactly one edge that points away from each vertex other than the root. To create the x_δ-tree T, we chose the least resistant edge exiting from each vertex other than x_δ. Moreover, each of these edges has a lower (or equal) resistance compared to the least-resistant edge exiting from x_δ. Hence there cannot be a rooted tree with lower total resistance than T. In fact, this argument shows that any x-tree has higher resistance than T unless x also maximizes $f_\delta(\cdot)$. From

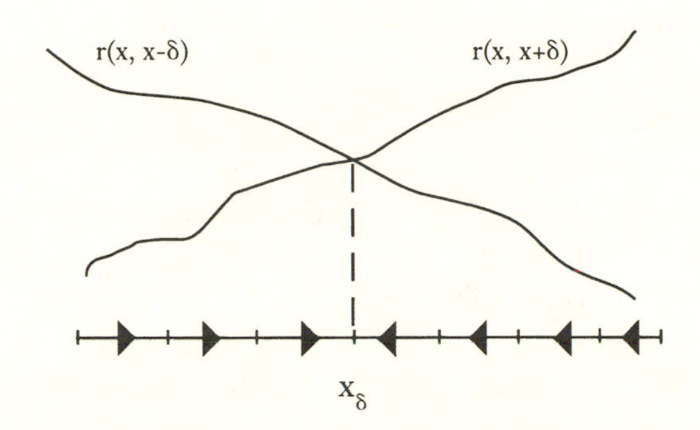

Figure 8.3. The least resistant rooted tree that is rooted at x_δ.

this and theorem 7.1, it follows that the stochastically stable state(s) are precisely those convention(s) h_x such that x maximize(s) $f_\delta(\cdot)$.

When δ is small and m is large relative to δ, any maximum x_δ of $f_\delta(x)$ lies close to the point x^* at which the curves $\delta amu'(x)/u(x)$ and $\delta bmv'(1-x)/v(1-x)$ intersect, namely,

$$au'(x^*)/u(x^*) = bv'(1-x^*)/v(1-x^*). \tag{8.3}$$

This is just the first-order condition for maximizing the concave function

$$a \ln u(x) + b \ln v(1-x), \tag{8.4}$$

which is equivalent to maximizing

$$u(x)^a v(1-x)^b. \tag{8.5}$$

The unique x^* that maximizes (8.5) is known as the *asymmetric Nash bargaining solution*. It follows that x_δ is arbitrarily close to the asymmetric Nash solution when δ is sufficiently small and m is sufficiently large relative to δ. This completes the outline of the argument.[1]

When errors are global, the proof is more complex. The reason is that certain large deviations may have low resistance. Suppose, in particular, that the process is currently in the norm h_x. Now suppose that a succession of row players erroneously make the weakest of all demands, δ. This induces the column player to switch to the best reply $(1-\delta)$

provided that the number of errors i satisfies $(i/bm)v(1-\delta) \geq v(1-x)$, that is,

$$i \geq \lceil bmv(1-x)/v(1-\delta) \rceil.$$

When x is close to 1 the right-hand side is small, that is, the transition from a norm h_x to the norm h_δ has a small resistance. Similarly, the transition from h_x (where x is close to zero) to the norm $h_{1-\delta}$ has a small resistance. Nevertheless, it can be shown that when δ is sufficiently small, this does not fundamentally change the conclusion—namely, that every least resistant x-tree is rooted at a maximum of $f_\delta(\cdot)$. The proof is given in Young (1993b).

Let us illustrate theorem 8.1 with a small example. Suppose that the row players sample one-third of the surviving records and have the utility function $u(x) = \sqrt{x}$. Suppose that the column players sample one-tenth of the surviving records and have the linear utility function $v(y) = y$. The asymmetric Nash solution is (5/8, 3/8), which optimizes (8.5). Let the precision be $\delta = .1$. When m is large, $f_\delta(x)/\delta m \approx \varphi_1(x) \wedge \varphi_2(x)$, where

$$\varphi_1(x) = (1/3)[1 - u(x - .1)/u(x)] = (1/3)(1 - \sqrt{1 - 1/10x}).$$

and

$$\varphi_2(x) = (1/10)[1 - v(1 - x - .1)/v(1 - x)] = 1/100(1 - x).$$

The function $\varphi_1(x) \wedge \varphi_2(x)$ is graphed in Figure 8.4. It achieves its maximum at a value between .6 and .7, and its maximum among the discrete demands occurs at 0.6. Hence the stochastically stable division is (.6, .4).

8.4 Variations on the Bargaining Model

How sensitive are the preceding results to the details of the model? A particularly important ingredient is the structure of the one-shot bargaining game. This game has the slightly unnatural feature that when players ask for too little—that is, when their demands sum to less than unity—the surplus is left on the table. Two natural alternatives suggest themselves. One is to suppose that, when the bargainers demand too little $(x + y < 1)$ they split the surplus $1 - x - y$ equally. The other

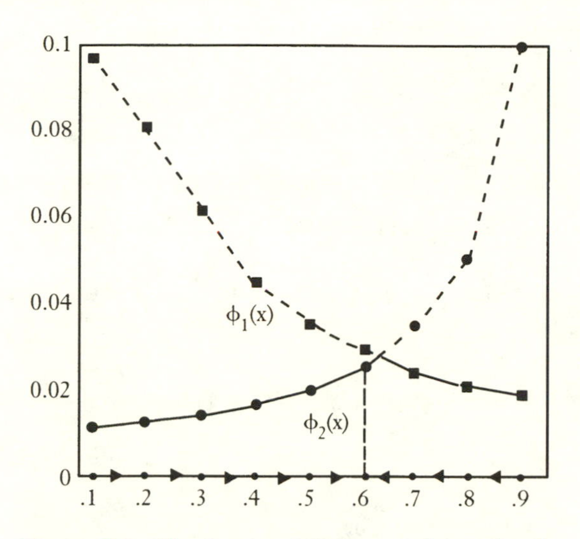

Figure 8.4. The least-resistant rooted tree for the situation $\delta = .1$, $u(x) = \sqrt{x}$, $v(1-x) = 1-x$.

approach is to suppose that they only conclude a deal if they coordinate exactly, that is, if $x + y = 1$. The consequences of the latter variation will be taken up in a more general setting in the next chapter. Here we shall briefly examine the first situation.

To keep the argument simple, let us assume that all errors are local and all agents have the same sample size. Suppose that the process is in the norm $(x, 1-x)$. The payoffs if the row player demands δ more than the norm, or the column player demands δ less than the norm, are as follows:

	$1-x$	$1-x-\delta$
x	$u(x), v(1-x)$	$u(x+\delta/2), v(1-x-\delta/2)$
$x+\delta$	$0, 0$	$u(x+\delta), v(1-x-\delta)$

Ignoring the sample size, the reduced resistance to moving from x to $x+\delta$ is

$$\tilde{r}(x, x+\delta) = \frac{u(x) - 0}{(u(x) - 0) + (u(x+\delta) - u(x+\delta/2))} \wedge \frac{v(1-x) - v(1-x-\delta/2)}{(v(1-x) - v(1-x-\delta/2)) + (v(1-x-\delta) - 0)}. \quad (8.6)$$

When δ is small, we have the approximations

$$u(x+\delta) - u(x+\delta/2) \approx (\delta/2)u'(x)$$

and

$$v(1-x) - v(1-x-\delta/2) \approx (\delta/2)v'(1-x).$$

From this we deduce that the first term in (8.6) is much larger than the second, and that

$$\tilde{r}(x, x+\delta) \approx (\delta/2)v'(1-x)/v(1-x).$$

Similarly,

$$\tilde{r}(x, x-\delta) = (\delta/2)u'(x)/u(x).$$

Arguing as before, we deduce that the stochastically stable norms are those that maximize $\tilde{r}(x, x-\delta) \wedge \tilde{r}(x, x+\delta)$, which are close to the value x^* such that $v'(1-x)/v(1-x) = u'(x)/u(x)$. This is the Nash bargaining solution. When the sample sizes differ, a similar argument leads to the asymmetric Nash solution.

8.5 Heterogeneous Populations

Adaptive models lend themselves naturally to the study of heterogeneous populations of agents, as we have already shown in Chapter 5. In the present context, bargainers can differ in at least two dimensions: degree of risk aversion and amount of information. We can therefore characterize each agent by a pair $\tau = (a, u)$ where $0 < a \leq 1/2$ is the fraction of records the agent samples, and $u(x)$ is a concave, strictly increasing utility function defined for all $x \in [0, 1]$. Without loss of generality we can assume that $u(0) = 0$ and $u(1) = 1$. The pair (a, u) represents the agent's *type*. Let T_1 be the set of all types represented among the row players, and let T_2 be the set of all types represented among the column players, where both T_1 and T_2 are finite.

As before, the essence of the problem is to find the least resistance to transiting from any given norm h_x to a neighboring norm $h_{x'}$ via local errors. (Indeed, the analysis can be carried out for a more general error structure, as shown in Young (1993b).) When δ is small and δm is large,

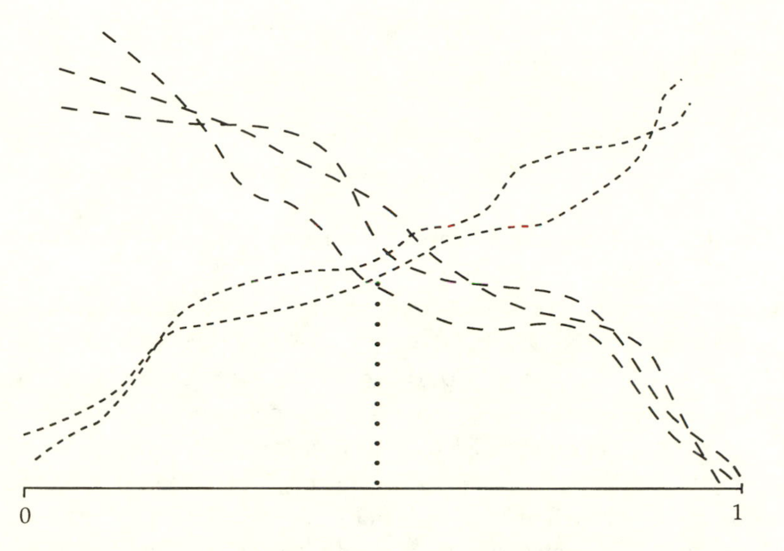

Figure 8.5. The stable outcome occurs where the lower envelope of the curves attains its maximum.

the resistances to the transitions $h_x \to h_{x-\delta}$ and $h_x \to h_{x+\delta}$ are given to a good approximation by the expressions

$$r(x, x-\delta) \approx \min_{(a,u)\in T_1} \{\delta amu'(x)/u(x)\}$$

and

$$r(x, x+\delta) \approx \min_{(b,v)\in T_2} \{\delta bmv'(1-x)/v(1-x)\}.$$

When δ is small and δm is large, the stochastically stable norms must be close to the unique point x^* where these curves cross. In other words, they are close to the point x^* where the lower envelope of the following family of curves achieves its maximum (see Figure 8.5 for an example):

$$\{au'(x)/u(x)\colon (a,u) \in T_1\} \cup \{bv'(1-x)/v(1-x)\colon (b,v) \in T_2\}. \tag{8.7}$$

Note that (8.7) has a unique maximum whenever the utility functions are concave, positive, and strictly increasing, which we have assumed they are.

THEOREM 8.2 (Young (1993b)). *Let G be the discrete Nash demand game with precision δ played adaptively by two finite populations consisting of types T_1*

and T_2, respectively. As δ becomes small, the stochastically stable division(s) converge to the unique maximum of

$$F_\delta(x) = \min_{\substack{(a,u)\in T_1 \\ (b,v)\in T_2}} \{au'(x)/u(x), bv'(1-x)/v(1-x)\}. \tag{8.8}$$

Consider the following example. The row population consists of two types of agents: ($a_1 = 1/4, u_1(x) = x$) and ($a_2 = 1/3, u_2(x) = x^{1/3}$). The column population consists of two other types of agents: ($b_1 = 1/5$, $v_1(y) = y$) and ($b_2 = 1/2, v_2(y) = y^{1/2}$). Among the row players, the criterion $au'(x)/u(x)$ is minimized by the second type of player for every x: $a_2u_2'(x)/u_2(x) = 1/9x$. Among the column players, the criterion $bv'(y)/v(y)$ is minimized by the first type of player for every y: $b_1v_1'(y)/v_1(y) = 1/5y$. Hence $F(x) = 1/9x \wedge 1/5(1-x)$. It achieves its unique maximum when $1/9x = 1/5(1-x)$, that is, when $x = 5/14$. In other words, the stochastically stable division can be made arbitrarily close to $(5/14, 9/14)$ for any two populations consisting of the types mentioned above, irrespective of the distribution of these types within each population.

8.6 Bargaining with Incomplete Information

The approach described above is conceptually quite different from the theory of bargaining under incomplete information, which is the standard way of treating heterogeneous populations (Harsanyi and Selten, 1972). In a bargaining model with incomplete information, two agents are drawn at random from their respective populations, and they play a noncooperative game. For simplicity we shall think of this as being the Nash demand game, though Harsanyi and Selten consider a more complicated noncooperative game. Each agent has a *belief* about the distribution of types in the opposite population, which may or may not correspond to the true distribution of types. An agent's *strategy* associates a demand with each possible type that he could be. Two strategies form a Bayesian equilibrium if each such demand maximizes the agent's expected payoff (contingent on his type) given his belief about the type of the other agent. Typically there will be a large number of such equilibria for a given set of beliefs. Harsanyi and Selten suggest a set of rational principles for selecting an equilibrium from this set, based on the idea of maximizing the product of the players' payoffs.

In the evolutionary approach, by contrast, agents need not know (or even have beliefs about) the distribution of types in the other population. Instead, they have beliefs about how much their opponent is likely to demand, which they infer empirically from samples of the other side's previous demands. A norm is a situation in which all members of the same population demand the same amount irrespective of their type, and all members of the opposite population demand the complementary amount irrespective of their type. Selection among the norms occurs not by a process of rational choice, but by an implicit process in which norms emerge and are displaced at different rates given the types that are present in each population.

8.7 Fifty-Fifty Division

Under some circumstances it is reasonable to expect that there will be some mixing between the row and column populations. In this case the same types will be present in the two populations, though they may not be present in the same frequencies. This idea can be modeled as follows. Let $T = T_1 = T_2$ be the set of all possible types that an agent could be. Let $\pi(\tau, \tau')$ be the probability that the type pair $(\tau, \tau') \in T \times T$ will be chosen to bargain in any given period. The first entry τ denotes the row player's type, and the second entry τ' denotes the column player's type. The distribution π *mixes roles* if every pair in $T \times T$ occurs with positive probability. Mixing is a natural consequence of mobility between classes. It also occurs if the classes are rigid but row and column players are drawn from the same "gene pool": in this case, every individual has a positive probability of being any type τ, though the probability of being a τ may differ for row and column.

When π mixes roles, theorem 8.2 holds as stated with $T_1 = T_2 = T$. Since the function $F(x)$ is symmetric about one-half, and since fifty-fifty is a feasible solution for all precisions δ, we obtain the following result.

Corollary 8.1. *Let G be the discrete Nash demand game with precision δ played adaptively by two finite populations in which the roles are mixed. For all sufficiently small δ the unique stochastically stable division is fifty-fifty.*

Recall that the usual argument for fifty-fifty division is that it constitutes a prominent focal point, which makes it an effective means of coordination (Schelling, 1960). An alternative argument is that fifty-fifty is perceived by many people to be fair, and people derive utility from treating others fairly. Together, these arguments say that fifty-fifty has value both as a coordination device and because it satisfies demands for fairness. However, in economic bargains where parties provide very different inputs (such as land and labor in agriculture) neither of these arguments is especially compelling. When the parties have asymmetric roles and are transparently *not* equal, fifty-fifty division loses prominence as a focal point and perhaps also its claim to being "fair." Nevertheless, fifty-fifty is in fact quite common in economic bargains between asymmetrically situated agents. In agriculture, for example, equal sharing of the crop is a very common form of contract in many parts of the world, including the midwestern United States.[2]

The above result provides a different explanation for fifty-fifty (and, more generally, for equal division of the surplus over and above the parties' disagreement outcomes). When bargainers are drawn from a given population of types, and their expectations are shaped by precedent, equal division is the most stable convention over time. Once established, it is the hardest to dislodge. Interpreted in this way, equal division may be a focal point *because* it is stable, not the other way around.

Chapter 9

CONTRACTS

THE PRECEDING chapter examined the effect of adaptive learning on a particular form of contract, namely, the division of a fixed pie. Here we shall extend the analysis to the evolution of contracts more generally. By a *contract*, we mean the terms that govern a relationship between people. Contracts may be written or unwritten, explicit or implicit. Some contracts are spelled out in fine detail, such as rental contracts between tenants and landlords or lending agreements between bankers and borrowers. Others are mostly implicit, such as the common understandings that a couple brings to a marriage. Still others occupy a middle ground: employment contracts are usually explicit about some matters, such as the number of hours to be worked, but quite vague about others, such as the relationship between performance and pay.

Whether the terms are implicit or explicit, what matters for the durability of a contract is that the parties know what is expected of them under various contingencies, and that behavior be consistent with expectations. This creates a demand for standard or conventional contracts that have been tested by experience, since the parties have a clearer idea about what their terms mean and how they will play out under different circumstances. Moreover, the existence of standard contracts makes it easier for the parties to come to terms in what would otherwise be a complex and perhaps indeterminate bargaining situation.

At the individual level we can model the choice of contract as a pure coordination game. The terms that a player demands are governed by his or her expectations about the terms that the other side will demand, and these expectations are governed by observations about the terms that agents in the other population actually did demand in previous periods. We further assume that the learning process is buffeted by small stochastic perturbations that represent idiosyncratic behavior and other minor disturbances. This learning process has quite striking implications for the welfare properties of contracts that are likely to be prevalent in the long run. Specifically, we shall show that adaptive learning tends to select contract forms that are efficient—the expected payoffs lie on the Pareto frontier. Furthermore, the payoffs from such contracts tend to be centrally located on the Pareto frontier instead of near the boundaries. When the payoffs form a discrete approximation of a convex set, the stochastically stable contract gives the parties approximately equal payoffs *relative* to the payoffs they could get under their most preferred contracts. Interpreted in the context of economic and social institutions, this says that the most stable contractual arrangements are those that are efficient, and more or less egalitarian, given the parties' payoff opportunities.

9.1 Choice of Contracts as a Coordination Game

Consider two disjoint classes of individuals—A and B—who can enter into relationships with each other. They might consist, for example, of employers and employees, men and women, owners and renters, or creditors and debtors. For simplicity, we shall assume that each relationship involves a pair of individuals, and that the terms of their relationship can be expressed in a finite number of alternative *contracts*. Let a_k be the expected payoff to the A-players from the kth contract, and b_k the expected payoff to the B-players, $1 \leq k \leq K$. We assume that these payoffs represent the players' choices under uncertainty, and that they satisfy the usual von Neumann–Morgenstern axioms. We shall also suppose that everyone knows his or her own payoff, but we need not suppose that anyone knows anyone else's payoff.

At the beginning of each period, a pair of players is drawn at random from $A \times B$, and each of them names one of the K contracts. If they name the same one, they enter into a relationship on those terms, and their expected payoffs are (a_k, b_k). If they name different contracts, they are unattached until the next matching. We shall assume that all contracts are *desirable*, that is, that the expected payoff from any feasible contract is strictly higher for both parties than the expected payoff from being unattached. Without loss of generality, we may normalize each player's utility function so that the payoff from the unattached state is zero. The recurrent game is therefore a $K \times K$ *pure coordination game* in which the diagonal payoffs are positive and the off-diagonal payoffs are zero. Indeed, the analysis below applies to any pure coordination game, whether or not it arises in this way.

Let $P^{m,s,\varepsilon}$ be adaptive learning applied to this situation. As usual, the error term reflects the idea that players sometimes make choices for idiosyncratic reasons that lie outside the model. Although these unconventional, quirky choices often result in missed opportunities with the more conventionally minded (and therefore quirky people are more likely to be unattached), if enough such choices accumulate, they can cause society to tip into a new norm. For the present, we shall assume (as in previous chapters) that these idiosyncratic shocks are independently and identically distributed among agents. In Section 9.7, we shall show that even when the shocks are correlated, the long-run outcome remains the same provided that the degree of correlation is not too large.

9.2 Maximin Contracts

A contract is *efficient* if there exists no other contract that yields higher payoffs to both parties. It is *strictly efficient* if there is no other contract that yields at least as high a payoff to one party and a strictly higher payoff to the other. Let $a^+ = \max_k a_k$ be the maximum payoff to the A-players under some contract; similarly let b^+ be the maximum payoff to the B-players. There is no loss of generality in scaling the utility functions so that $a^+ = b^+ = 1$, and we shall assume henceforth that this normalization has been made. Define the *welfare index* of contract k to be the smaller of a_k and b_k,

$$w_k = a_k \wedge b_k, \tag{9.1}$$

and let

$$w^+ = \max_k w_k. \tag{9.2}$$

A contract with welfare index w^+ will be called a *maximin contract*. A maximin contract favors the least-favored class in the sense that no other contract yields a higher expected payoff to the class that is least well-off relative to its most-preferred contract.

Figure 9.1 illustrates the welfare indices of eight different contracts, together with the maximin contract. The intersection of the dotted lines with the diagonal determine the welfare indices, and w^+ is the value that lies furthest out along the diagonal.

Why, though, is it meaningful to compare von Neumann–Morgenstern utilities in this way? From the individual's perspective, it is not meaningful, but we do not assume that individuals make such comparisons. (How could they if they do not know the others' utility functions?) The claim is that *society* makes interpersonal comparisons through the feedback effect of precedents on expectations. Moreover, there is a simple

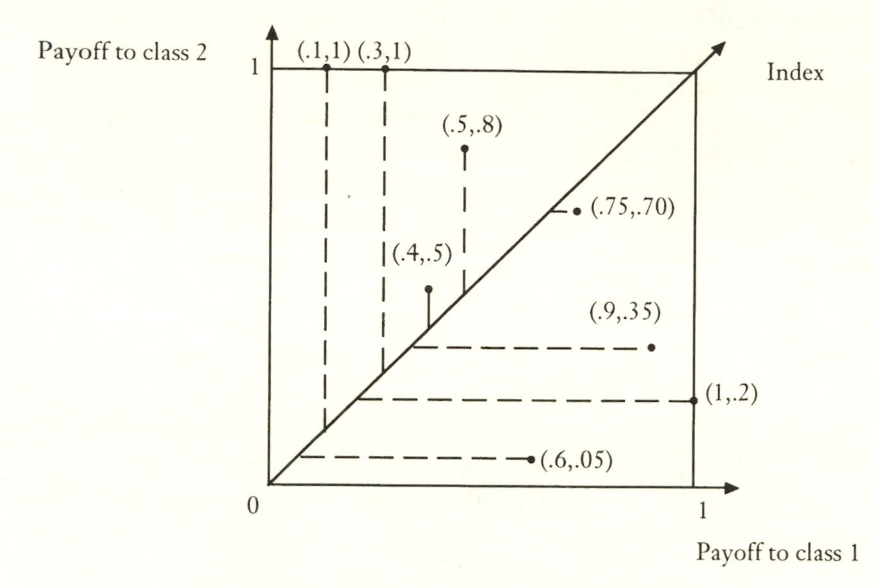

Figure 9.1. Welfare indices for eight contracts.

explanation for this phenomenon: individuals choose actions based on their payoffs times the probabilities with which they expect the *other* side to choose these same actions. In a stochastic environment, this means that individuals with different utility functions adjust their choice behavior at different rates. Interpersonal comparisons therefore arise implicitly because differences in von Neumann–Morgenstern utility imply different rates of behavioral adaptation in a stochastic environment. At the societal level, this creates a selection bias in favor of the maximin norm, or something close to it, as we shall now show.

9.3 A Contract Selection Theorem

To formulate our result precisely, we need some further notation. Let a^- denote the lowest payoff to the A players among all contracts in which the B players get their maximum, and define b^- similarly (see Figure 9.1). Let

$$w^- = a^- \vee b^-. \tag{9.3}$$

Normally we would expect w^- to be small, because eking out the maximum possible gain for one class will generally occur at the expense of the other class as a result of substitution possibilities. We claim that the smaller w^- is, the closer is the stochastically stable outcome to the maximin solution. Specifically, define the *distortion parameter* α as

$$\alpha = w^-(1-(w^+)^2)/(1+w^-)(w^+ + w^-). \tag{9.4}$$

Note that α is small when w^- is small and/or w^+ is close to one.

THEOREM 9.1. *Let G be a two-person pure coordination game, and let $P^{m,s,\varepsilon}$ be adaptive play. If s/m is sufficiently small, every stochastically stable state is a convention; moreover, if s is sufficiently large, every such convention is efficient and approximately maximin in the sense that its welfare index is at least* $(w^+ - \alpha)/(1+\alpha)$.

The usefulness of this lower bound depends on the problem at hand. In the example shown in Figure 9.1, $w^+ = .700$ and $w^- = .200$, so $\alpha = .067$. The theorem therefore says that the stochastically stable outcome has welfare index at least $(.700 - .067)/1.067 = .593$. It is evident from the figure that the only contract that satisfies this criterion is the maximin contract. If there were other contracts with welfare index between .593 and .700, the theorem would not eliminate them; in general, however, the addition of more contracts will tend to increase w^+ and to decrease w^-, which reduces α and correspondingly increases the cutting power of the theorem.

While theorem 9.1 does not provide the sharpest possible lower bound in all coordination games, it is best possible for some games, as we show by example in Section 9.5. There are important classes of games, moreover, for which the stochastically stable outcome is essentially the same as the maximin solution. These include 2×2 coordination games, symmetric coordination games, and coordination games whose payoffs form a discrete approximation of a convex bargaining set. Indeed, in the latter case the stochastically stable outcome is the Kalai-Smorodinsky solution, as we show in Section 9.7.

The general approach to computing the stochastically stable outcome relies on the methods developed in previous chapters. Specifically, equation (7.2) shows that the resistance to transiting from coordination equilibrium j to coordination equilibrium k (as a function of the sample size s) is given by the expression

$$r_{jk}^{s} = \lceil s r_{jk} \rceil, \text{ where } r_{jk} = a_j/(a_j + a_k) \wedge b_j/(b_j + b_k). \quad (9.5)$$

We then apply the method of rooted trees to determine the equilibrium with minimum stochastic potential, as described in Chapter 3.

9.4 The Marriage Game

Let us illustrate these ideas with an example involving marriage contracts. It is a commonplace observation that men and women often play different roles in a marriage, and that these differences depend to a substantial extent on social custom. Marital arrangements that are normal and customary in Papua New Guinea may not be normal or customary in Sweden, for example. Furthermore, social expectations about these matters change over time. In Europe and the United States, for example, the rights of married women have evolved markedly over the past two hundred years from a situation where they enjoyed very little autonomy to one that more closely approximates equality. This change resulted from gradual shifts in attitudes and expectations in the general population, rather than from any single defining event; hence it has a distinctly evolutionary flavor. While the theory described above is not intended to explain such historical developments in detail, it can nevertheless shed light on a related question: in partnerships between people who are complementary in some respect, what distribution of rights within the partnership is the most stable over the long run?

To analyze this problem, consider the following "marriage game." There are two classes of agents—say men and women—who can enter into partnerships or "marriages" with one another. Each marriage is governed by one of three types of contractual arrangements: the man controls, the woman controls, or they share control. Let us assume that each of these arrangements is equally efficient, so that the only issue is the distribution of power in the relationship. Let us further assume that the likelihood that an individual enters into a given contractual arrangement is influenced by the degree to which that arrangement is *customary* in the given society.

The evolutionary process operates as follows. In each period, one man and one woman are tentatively matched, and each proposes one of the three possible contracts. If they name the same contract they get married; if they name different ones they break up. Suppose the payoffs are as follows:

The Marriage Game

		Men: Take Control	Men: Share Control	Men: Cede Control
	Take Control	0, 0	0, 0	5, 1
Women	Share Control	0, 0	3, 3	0, 0
	Cede Control	1, 5	0, 0	0, 0

The fact that each outcome has the same total utility is not significant here; indeed the utility functions can be re-scaled in any way that we wish without affecting the resistances between the three coordination equilibria (see equation (9.5)). These resistances are illustrated in Figure 9.2. From this we see that the sharing norm is the unique stochastically stable outcome whenever the sample size is sufficiently large. Concretely this means that, when the noise is small, the sharing norm will be observed with substantially higher probability over the long run than will either of the other arrangements.

It is important to recognize that this result depends on the *welfare* implications of the various contractual arrangements, not on their specific *terms*. In particular, the model does not necessarily predict equal treatment in situations where the parties have unequal abilities or tastes. In this case, efficiency might well call for asymmetric arrangements in which each partner takes on those tasks to which he or she is particularly inclined or particularly well-suited. The general point of theorem 9.1 is that there is a long-run selection bias toward contractual arrangements that offer both sides a fairly high level of welfare *relative to* the welfare they could enjoy under alternative arrangements.

Moreover, there is a simple intuitive explanation for this result. Conventions with extreme payoff implications are relatively easy to dislodge because the members of one of the groups are dissatisfied compared to what they could get under some other arrangement. It does not take many stochastic shocks to create an environment in which members of the dissatisfied group prefer to try something different. This involves a risk of being "unattached," but if the payoffs from being attached are sufficiently meager, it may be a risk worth taking. Under the sharing norm, by contrast, both sides enjoy reasonably high payoffs given their

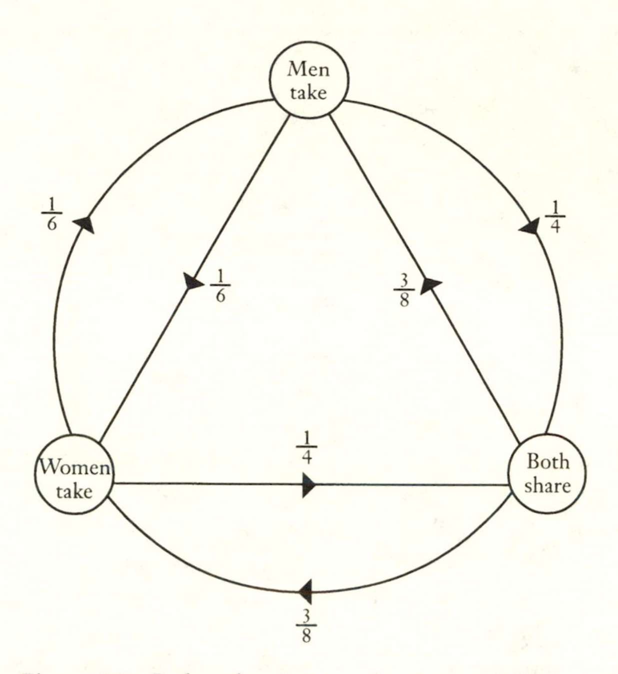

Figure 9.2. Reduced resistances between equilibria in the marriage game.

alternatives; hence there is less impetus for change. More generally, theorem 9.1 says that the evolutionary process tends over time to push society away from the boundaries of the feasible payoff set, and move it toward the "middle" of the efficiency frontier, as defined by the maximin criterion.

9.5 Examples Showing Departures from Strict Efficiency and Exact Maximin

We now show why the theorem cannot be substantially strengthened. First, we exhibit a situation where the (stochastically) stable contract is efficient but not strictly so. In Figure 9.3, contract 1 weakly Pareto-dominates contract 2, but the least resistant 1-tree (left panel) and the least resistant 2-tree (right panel) have the same resistance. Hence both contracts are stable.

The next example shows that the bound in the theorem is tight when there are at least four games. Choose real numbers w^-, w, and w^+ such

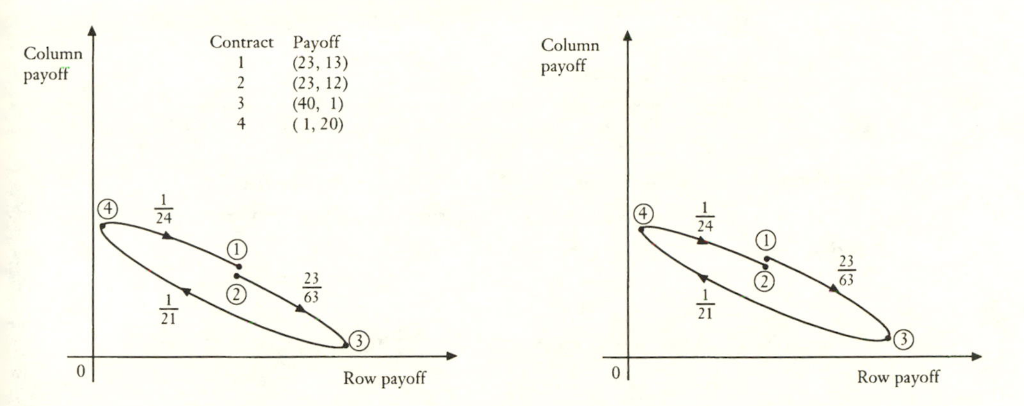

Figure 9.3. Left: Least resistant 1-tree. Right: Least resistant 2-tree.

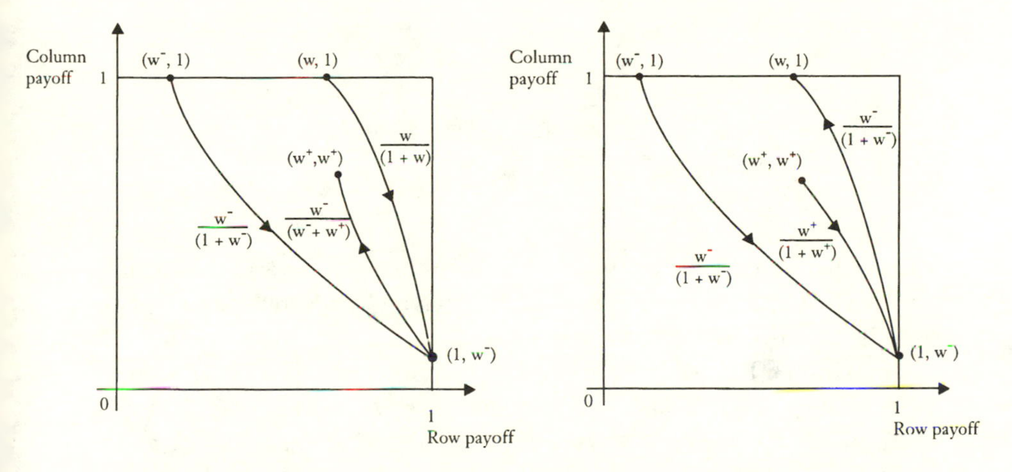

Figure 9.4. Left: Least resistant (w^+, w^+)–tree. Right: Least resistant $(w, 1)$–tree.

that

$$0 < w^- < w < w^+ < 1.$$

Consider four contracts with the payoffs shown in Figure 9.4, where w^- and w^+ have the meanings defined in (9.2) and (9.3). The left panel shows the least resistant tree rooted at (w^+, w^+); the right panel shows the least resistant tree rooted at $(w, 1)$. The contract with payoffs $(w, 1)$ has the same or lower stochastic potential than the maximin contract

(w^+, w^+) if

$$\frac{w^-}{1+w^-} + \frac{w^+}{1+w^+} + \frac{w^-}{1+w^-} \leq \frac{w^-}{1+w^-} + \frac{w}{1+w} + \frac{w^-}{w^- + w^+}.$$

This reduces to $w \geq (w^+ - \alpha)/(1 + \alpha)$, where α is defined as in (9.4). Hence the bound in theorem 9.1 is best possible when the game is 4×4, and the argument extends naturally to larger games.

9.6 Small Games and Symmetric Coordination Games

In this section we discuss two special classes of games where there is no distortion in the selection process, and the maximin outcome is stochastically stable. One such class consists of 2×2 coordination games, where the payoffs take the form

	1	2
1	a_1, b_1	$0, 0$
2	$0, 0$	a_2, b_2

$a_1, b_1, a_2, b_2 > 0.$

From theorem 4.1 we know that adaptive learning selects the equilibria that maximize the product of the players' expected payoffs, that is, the risk dominant equilibria. We claim that any such equilibrium is maximin. Let us assume that $(1, 1)$ risk dominates $(2, 2)$, that is, $a_1 b_1 \geq a_2 b_2$. If (1, 1) also Pareto dominates $(2, 2)$, then $(1, 1)$ has welfare index $w_1 = a_1/a_1 \wedge b_1/b_1 = 1$ whereas $(2, 2)$ has welfare index $w_2 = a_2/a_1 \wedge b_2/b_1 < 1$. Hence the former is the maximin equilibrium. If $(1, 1)$ does not Pareto dominate $(2, 2)$, we may suppose that $a_1 < a_2$ and $b_1 \geq b_2$. Thus the welfare index of $(1, 1)$ is $w_1 = a_1/a_2 \wedge b_1/b_1 = a_1/a_2$, whereas the welfare index of $(2, 2)$ is $w_2 = a_2/a_2 \wedge b_2/b_1 = b_2/b_1$. Since $a_1/a_2 \geq b_2/b_1$, equilibrium $(1, 1)$ is maximin.

Next, consider the case of a symmetric, two-person coordination game, where $a_k = b_k$ for all k. This situation was first analyzed by Kandori and Rob (1995). Without loss of generality we can index the equilibria in decreasing order of welfare: $a_1 \geq a_2 \geq \cdots \geq a_K$. The Pareto frontier consists of the payoff pair (a_1, a_1), and the assertion is that the stochastically stable outcomes are precisely those k such that $a_k = a_1$.

This result can be established quite simply as follows. Assume for simplicity that $a_1 > a_k$ for all $k \geq 2$. Let us also assume that s/m

is sufficiently small that the conventions are the only recurrent states of the unperturbed process. Represent each such state by a vertex of a graph, and label the vertices $1, 2, \ldots, K$. By (9.5), the minimum resistance edge exiting from vertex $k > 1$ is directed toward vertex 1, and its reduced resistance is strictly less than 1/2. On the other hand, every edge exiting from vertex 1 has reduced resistance at least 1/2. It follows that the tree rooted at vertex 1 and consisting of the directed edges $\{(2, 1), (3, 1), \ldots, (K, 1)\}$ has a strictly smaller reduced resistance than any other rooted tree. Hence it has minimum resistance among all rooted trees when the sample size s is sufficiently large. Thus equilibrium 1 corresponds to the unique stochastically stable convention.

9.7 The Kalai-Smorodinsky Solution

Suppose that the payoffs from the various contracts lie in a compact, convex set $C \subseteq R^2_+$. Suppose also that C is *comprehensive*: given any $(a, b) \in C$, C contains all those pairs (a', b') satisfying $(0, 0) \leq (a', b') \leq (a, b)$. Assume finally that C contains a strictly positive payoff pair. Then C is a *bargaining set*, that is, a compact, convex, comprehensive, full-dimensional subset of R^2_+.

Given any such bargaining set C, let $a^+ = \max\{a : (a, b) \in C\}$ and $b^+ = \max\{b : (a, b) \in C\}$. The *Kalai-Smorodinsky solution* is the unique vector (a^*, b^*) in C such that $a^*/a^+ = b^*/b^+$ is a maximum. Denote this maximum value by w^+.

Discretize C as follows: for each small $\delta > 0$, let C_δ consist of all payoff vectors $(a, b) \in C$ such that a/δ and b/δ are strictly positive integers. Evidently, $C_\delta \to C$ in the Hausdorff metric. Rescaling the utility functions if necessary, define $w^-(\delta)$ and $w^+(\delta)$ as in (9.2) and (9.3). Since C is comprehensive, $w^-(\delta) \to 0$ as $\delta \to 0$. Since C is convex, $w^+(\delta) \to w^+$ as $\delta \to 0$. Defining $\alpha(\delta)$ as in (9.4), it follows that $\alpha(\delta) \to 0$ as $\delta \to 0$. Hence the stochastically stable conventions guaranteed by theorem 9.1 must be close to the Kalai-Smorodinsky solution when δ is small.

To formulate this result precisely, let us say that a pair $(a, b) \in C_\delta$ is *stochastically stable* if the associated convention is stochastically stable in $P^{m,s,\varepsilon}$ for all sufficiently large s and all sufficiently small s/m.

THEOREM 9.2. *Let C be a bargaining set and let G_δ be the pure coordination game with coordination equilibrium payoffs in C_δ for every small precision*

δ. As δ becomes small, the stochastically stable payoffs converge to the Kalai-Smorodinsky solution of C.

9.8 Correlated Shocks

In this and preceding chapters, we have modeled stochastic variation by supposing that changes in behavior are driven by the accumulation of many uncoordinated, idiosyncratic choices. Of course, this is not the only way in which social change occurs. Sometimes individuals react to a common event, such as a technology shock. Or their expectations might be changed by the actions of a highly visible person (a role model). Without trying to downplay the variety and complexity of such stochastic influences, we can nevertheless say that they do not, by themselves, change the substance of the argument. To see why, consider a coordination game in which the current norm is h_j. To tip the process into the basin of attraction of some other norm h_k requires that at least $\lceil sr_{jk} \rceil$ people demand regime k instead of regime j. When these changes arise from uncorrelated aberrations in behavior, each having probability ε, the probability of this event is on the order of $\varepsilon^{\lceil sr_{jk} \rceil}$.

Now suppose instead that individuals do not change idiosyncratically, but in groups. To be concrete, suppose that whenever an individual makes an idiosyncratic choice, then for the next $p - 1$ periods it is certain that everyone in the same class will copy this choice. This corresponds to the notion that new ideas affect everyone in a given class, but that after p periods, the idea wears off. (An example would be a book that extols the virtue of regime k over regime j.) Assume that the probability that a new idea will hit a given class in any given period is ε. In this correlated version of the model, it takes only $\lceil sr_{jk}/p \rceil$ "hits," each p periods apart, to tip the process from h_j to h_k. When ε is small, the probability of this event is on the order of $\varepsilon^{\lceil sr_{jk}/p \rceil}$. It follows that the stochastically stable norms in both the correlated and uncorrelated models minimize the same stochastic potential function (based on the reduced resistances r_{jk}) whenever s is sufficiently large relative to p.

While the framework described above is admittedly a stylized model of contract formation, it does contain some of the key elements that are surely relevant. These include the salience of precedent in shaping expectations, boundedly rational responses by individuals to complex environments, and idiosyncratic variation in behaviors. The precise way in which these features are modeled may alter the conclusions to some degree. Nevertheless it seems reasonable to conjecture that the most likely outcomes of such a process will tend to lie toward the "middle" of the feasible payoff set. The reason is that contractual relationships whose payoffs lie near the boundary tend to be unstable. They imply that some group is dissatisfied, and the more dissatisfied a group is, the more easily it is seduced by new ideas that give its members hope of getting more. Change, in other words, is driven by those who have the most to gain from change. Over the long run, this tends to favor contracts that are efficient and that offer each side fairly high payoffs within the set of payoffs that are possible.

Chapter 10

CONCLUSION

THE THEORY we have sketched has two general implications. On the one hand, it demonstrates how high-rationality solution concepts in game theory can emerge in a world populated by low-rationality agents. Among the concepts we recover via this route are the Nash bargaining solution, the risk-dominant equilibrium in 2×2 games, iterated elimination of dominated strategies, minimal curb sets, and efficient equilibria in games of pure coordination. Moreover, in some types of extensive-form games, one obtains subgame perfect equilibrium and various forms of forward induction (Nöldeke and Samuelson, 1993). Interpreted in this way, the evolutionary approach is a means of reconstructing game theory with minimal requirements about knowledge and rationality.

Interpreted more broadly, the theory suggests how complex economic and social structure can emerge from the simple, uncoordinated actions of many individuals. When an interaction occurs over and over again and involves a changing cast of characters, a feedback loop is established whereby past experiences of some agents shape the current expectations of other agents. This process yields predictable patterns of equilibrium and disequilibrium behavior that can be construed as social and economic "institutions," that is, established customs, usages, norms, and forms of organization. Although we have taken games as our basic model of interaction, the theory can be applied to many other forms of interaction, as we illustrated with the neighborhood segregation model.

Admittedly, there is a considerable gap between the simple models of interaction that we have studied and the economic and social institutions that we see around us. One would be hard-pressed, for example, to identify the complete set of rules and incentives that govern the negotiation and enforcement of economic contracts. It would be even more difficult to write down the game that represents, say, interactions in the workplace or the family. Nevertheless, these institutions can be thought of as equilibria in appropriately defined, high-dimensional games. The same goes for conventions of correct or morally acceptable behavior. We sometimes refer to these as "norms" instead of "conventions," thus expressing the idea that deviations from the norm may be punished. For our purposes, however, this is not a fundamental distinction: norms can also be represented as equilibria in a repeated game, where social opprobrium and other forms of punishment are the expected responses to deviations from the norm (and those who fail to carry out the appropriate punishments may themselves be punished). Our theory applies to all of these situations: people develop expectations about how others behave through repeated interactions and experience of others' interactions, which eventually coalesce into recognizable patterns of behavior.

Even if we grant, however, that customs and norms can be thought of as equilibria in games, is it true that they arise through the accretion of many uncoordinated decisions, or do they arise through the concerted and deliberate action of a few key people? Obviously it would be absurd to claim that they arise *only* in the former manner. Many institutions and patterns of behavior have been shaped, at one time or another, by influential people who endorsed (or enforced) a particular way of doing things. Napoleon instituted the legal code that still governs much of continental Europe; Bismarck established a social security system for industrial workers in Germany that served as a model for many later systems;[1] Catherine de Medici made it fashionable in France to use a fork at dinner.[2] Major players obviously matter in the development of economic and social institutions, but this does not imply that minor players do *not* matter. Actions by major players stand out and are easy to identify; small variations in individual behavior are more subtle and difficult to pinpoint, but may ultimately be more important for the development of some kinds of institutions. Moreover, we suspect that influential actors often get credit for things that were about to happen anyway.

Even if major players do sometimes matter, they may be minor relative to the scale of the social institution under consideration. Consider the evolution of language: Is it governed by major or minor players? Words become current in part through street talk, in part through education, and in part through dissemination by the media. It is hard to say which matters more over the long run. It should also be remembered that coordinated decisions by large groups of players are often quite small in the overall scheme of things. Consider the teaching of a second language, such as Chinese or Japanese, in the United States. The choice of which languages to offer are usually made by schools rather than by individuals, so we might think that the choice is governed by "large" players. The fact is, however, that there are many schools, and they often make choices in an uncoordinated way. Schools in Nebraska probably do not coordinate their language curricula with schools in Iowa. Even if they did, they would almost certainly not coordinate their choice with schools in Germany. At almost any level we can conceive of, these decisions are made in a fairly decentralized way relative to the whole. Moreover, it is the whole that matters: the value of learning a second language depends on the number of other people in the world who learn it. The players in this coordination game can be very large indeed (countries) and still be quite small on the scale of the whole process.

A similar argument can be made for the evolution of many other types of institutions: the adoption of currencies, the use of particular kinds of contracts (e.g., rental contracts, employment contracts, marriage contracts), codes of socially acceptable behavior, acceptable punishments for deviations from socially acceptable behavior, and so forth. All such forms of interaction are sustained as institutions through the mutually consistent expectations of society's members. Change is driven in part by small individual variations that tip expectations into a new equilibrium, and in part by the concerted actions of influential individuals and groups. We have emphasized the role played by the small players, while not denying the importance of the larger ones.

Another important issue is the *rate* at which change occurs. Change is driven in part by the underlying dynamics of adjustment and in part by idiosyncratic shocks. The former will typically operate more quickly than the latter. In practice, this means that the trajectory of the process in the short run will be strongly influenced by initial conditions, and the dynamics will be fairly well approximated by the *expected* motion. Idiosyncratic shocks will be felt only over the longer run, as various regimes become established and are later undone. The length of time needed for these regime transitions depends on the size of the stochastic shocks and the degree of correlation between them, the amount of information agents use in making their decisions, and the extent to which they interact in small, close-knit groups. It should also be recalled that the waiting time in these models is measured in "event" time, and thousands or even millions of distinct events may be compressed into a short interval of "real" time. Thus the length of the long run depends crucially on the details of the learning environment. In general, however, it should come as no surprise to find societies that operate for long periods of time in regimes that do *not* correspond with the long-run predictions of the theory. Institutions, once established, can lock people into established ways of thinking that are hard to undo. Nevertheless established institutions can be undone, and over time they are.

The preceding chapters have outlined a general approach to studying the dynamics of institutional change, based on a simple model of decentralized agents who adapt their expectations and responses to a changing environment. We assumed in particular that people form simple statistical models to predict how others will behave in a given situation. Usually (but not always) they choose myopic best replies given their expectations. Obviously many variants and elaborations of such a learning process can be imagined, some of which were discussed in Chapter 2. It is natural to ask, for example, how sensitive the results are to different ways of specifying the learning rule. It would be particularly interesting to examine the behavior of a population composed of different types of learners: some who employ statistical forecasts, others who imitate, and still others who respond in Pavlovian fashion to payoffs from past experience.

Another elaboration of the model would recognize that interactions between players are to some extent endogenous.[3] The probability that two players will meet may depend on the history of the process to date—on the number of times they met before, on the payoffs that resulted from their previous interactions, and so forth. In a model where individuals trade goods, for example, large gains from trade might well reinforce the probability that agents will trade in the future. This would lead to the endogenous formation of trading networks. In situations where people interact socially, one could imagine that there would be a tendency for people to select partners who are like themselves along such dimensions as ethnicity, income, language, and geographical location. One could further suppose that the fact of interacting makes them even more similar. Such a process can congeal into distinct tribes or castes that have many internal interactions but only limited contact with each other.[4] In the presence of repeated shocks, one should be able to predict the relative likelihood that stratified (or integrated) configurations will emerge over the long run. The point of this study has been to sketch a general framework for tackling these questions, rather than to argue that any single learning rule is correct, or that any one model of social interaction is more realistic than another.

There are, nevertheless, several qualitative features of such processes that are likely to be robust under different model specifications and that are subject (at least in principle) to empirical verification. To illustrate these features concretely, imagine a population divided into many similar subpopulations, or "villages," that do not interact with one another. (They might be located on different islands, for example.) Consider a form of interaction—a game—that is played recurrently in each of the villages, and suppose that this game has multiple strict Nash equilibria. For example, the interaction might involve a principal and an agent who are trying to coordinate on the form of the contract that will govern their relationship. Suppose further that the system is subjected to persistent idiosyncratic shocks that are small but not vanishingly small. The following three features are valid for the adaptive dynamics analyzed here, and no doubt hold for a considerably wider class of stochastic learning processes.

Local conformity, global diversity. Within each village, at any given time, behaviors are likely to be close to some equilibrium (the local convention), though idiosyncratic, unconventional behaviors will be present to some degree. Different villages may operate with different conventions due to historical chance.

Punctuated equilibrium. Within each village, there will tend to be long periods of stasis in which a given equilibrium remains in place, punctuated by occasional episodes in which the village tips from one equilibrium to another in response to stochastic shocks. Thus there is temporal diversity in a given place as well as spatial diversity at a given time.

Equilibrium stability. Some equilibria are inherently more stable than others, and, once established, they tend to persist for longer periods of time. The relative stability of different equilibria will tend to be reflected in the frequency with which they occur among the various villages at a given point in time (and also within each village over a long period of time). When the stochastic perturbations are small, the mode of this frequency distribution will be close to the stochastically stable equilibrium with high probability.

The challenge to theory is to determine whether these and related qualitative predictions are borne out by empirical evidence. In looking for evidence, one would want to focus on games that are played frequently by changing partners, in which precedent plays an important role in shaping expectations, and in which there are positive reinforcement effects from conformity (as in a coordination game). Forms of contracts are natural candidates because their enforcement is carried out by courts, which tend to rely on precedent to determine the meaning of contract terms and the remedies for default. Standards of professional practice (in medicine, accounting, or law) may lend themselves to a similar type of analysis. Means of communication, including the evolution of language, also have many of the relevant characteristics. The reader will no doubt think of other examples.

Of course, the theory developed here, like all theories, abstracts away from complications that inevitably arise in applications. One complication is that games change over time. If they change too rapidly, the adaptive learning process may not have time to settle into its long-run pattern. In these situations the transient rather than the asymptotic behavior of the adaptive process should occupy center stage in the analysis. A second complication is that games, and the conventions they generate, cannot always be treated in isolation. Almost every game we can think of is embedded within a larger game. Bargaining over contract forms takes place within the shadow of the law, the law operates within the penumbra of morality, morality is colored by religious belief. Norms and customs from one sphere inevitably spill over into different spheres; there are no clear boundaries delimiting our relations with others. Doubtless all of these interactions could be written down as one large game. Equally doubtless they never will be.

Still, we must begin somewhere. Just as perfect spheres and frictionless planes are idealized but useful concepts for modelling mechanical interactions, so we can view games and learning rules as primitives for modelling social and economic interactions. While our approach cannot be expected to predict the history of any single institutional form, it does suggest that institutions evolve according to specific spatial and temporal patterns, and that there is a relationship between the durability of institutions and their welfare implications for individuals.

Appendix

PROOFS OF SELECTED THEOREMS

LEMMA 3.1 Freidlin and Wentzell, 1984). *Let P be an irreducible Markov process defined o ι a finite state space Z. The stationary distribution μ of P has the property that the probability of each state z is proportional to the sum of the likelihoods of its z-trees, that is,*

$$\mu(z) = \nu(z) / \sum_{w \in Z} \nu(w), \text{ where } \nu(z) = \sum_{T \in \mathcal{T}_z} P(T). \tag{A.1}$$

PROOF. Let us view the states as vertices of a graph. For any two distinct states z and z' denote the edge directed from z to z' by the ordered pair (z, z'). Fix a particular state z. A *z-cycle* is a subset C of directed edges such that: (i) C contains a single directed cycle, which involves z, and (ii) for every state w not in the cycle there is a unique directed path in C from w to the cycle. See Figure A.1 for an illustration of this concept. Let C_z denote the set of all z-cycles. Recall that a w-tree T is a set of directed edges such that from every vertex $w' \neq w$ there is a unique directed path from w' to w. Observe that every z-cycle C can be written in the form $C = T \cup \{(w, z)\}$, where T is a w-tree and $w \neq z$. Similarly, every z-cycle can be written in the form $C = T \cup \{(z, w)\}$, where T is a z-tree and $w \neq z$. We shall exploit these equivalent representations below.

Let P be an irreducible Markov process on Z, where $P_{zz'}$ denotes the probability of transiting from z to z' in one period. For any subset S of directed edges, define

$$P(S) = \prod_{(z,z') \in S} P_{zz'}.$$

For each $z \in Z$ let

$$\nu(z) = \sum_{T \in \mathcal{T}_z} P(T).$$

Note that $\nu(z) > 0$ because the process is irreducible. By virtue of the connection between z-cycles and w-trees remarked on above, we have

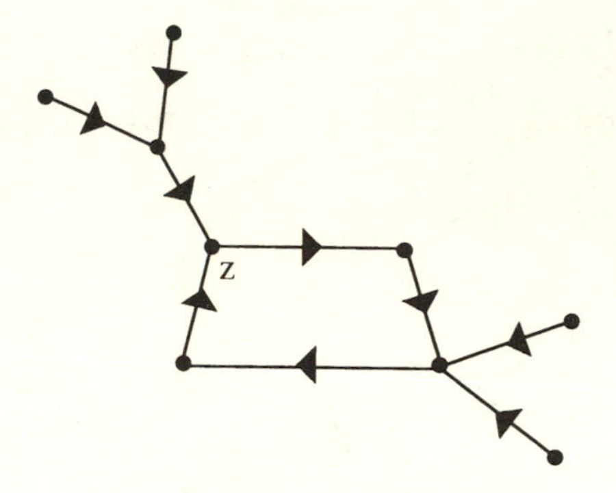

Figure A.1. Illustration of a z-cycle.

the identities

$$\sum_{C \in \mathcal{C}_z} P(C) = \sum_{w:w \neq z} \nu(w) P_{wz}, \tag{A.2}$$

$$\sum_{C \in \mathcal{C}_z} P(C) = \nu(z) \left(\sum_{w:w \neq z} P_{zw} \right). \tag{A.3}$$

It follows that

$$\sum_{w:w \neq z} \nu(w) P_{wz} = \nu(z) \left(\sum_{w:w \neq z} P_{zw} \right). \tag{A.4}$$

We also have the identity

$$\nu(z) \left(\sum_{w:w \neq z} P_{zw} \right) = \nu(z)(1 - P_{zz}). \tag{A.5}$$

Combining (A.4) and (A.5) we deduce that

$$\sum_{w \in Z} \nu(w) P_{wz} = \nu(z). \tag{A.6}$$

In other words, ν satisfies the stationarity equation for P. Hence its normalization μ does also, so μ is the stationary distribution of P. This concludes the proof of lemma 3.1.

THEOREM 3.1. *Let P^ε be a regular perturbed Markov process and let μ^ε be the unique stationary distribution of P^ε for each $\varepsilon > 0$. Then $\lim_{\varepsilon \to 0} \mu^\varepsilon = \mu^0$ exists, and μ^0 is a stationary distribution of P^0. The stochastically stable states are precisely those states that are contained in the recurrent class(es) of P^0 having minimum stochastic potential.*

PROOF. Let P^ε be a regular, perturbed Markov process defined for all $\varepsilon \in [0, \varepsilon^*]$, and denote the recurrence classes of P^0 by $E_1, E_2, \ldots, E_K$. Construct a graph Γ^0 having vertices $1, 2, \ldots, K$, where the jth vertex corresponds to the class E_j. Given two vertices $1 \leq i \neq j \leq K$, let r_{ij} denote the resistance of the least resistant path in Z that begins in E_i and ends in E_j. Recall that, for each j, the *stochastic potential* γ_j of the class E_j is defined to be the resistance of the least resistant j-tree in Γ^0. To prove the theorem, we need to show that $\lim_{\varepsilon \to 0} \mu^\varepsilon = \mu^0$ exists, and that $\mu^0(z) > 0$ if and only if $z \in E_j$ where j minimizes γ_j.

We begin by characterizing the stochastically stable classes (and states) in terms of rooted trees whose vertex set is the entire space Z. Since P^ε is a regular perturbation, for every pair of distinct states z and z' there exists a number $r(z, z')$ such that $P^\varepsilon_{zz'}$ converges to $P^0_{zz'}$ at a rate on the order of $\varepsilon^{r(z,z')}$. (See expression 3.12.) We adopt the convention that $r(z, z') = \infty$ if $P^\varepsilon_{zz'}$ is identically zero on the interval $[0, \varepsilon^*]$. Let Γ^* be the complete directed graph whose vertex set is Z such that each directed edge (z, z') has weight $r(z, z')$. Let $\mathcal{T}^*_z$ denote the set of all z-trees on Z. Define the function $\gamma\colon Z \to R$ as follows:

$$\gamma(z) = \min_{T^* \in \mathcal{T}^*_z} \sum_{(z',z'') \in T^*} r(z', z''). \tag{A.7}$$

Note that, since P^ε is irreducible for each $\varepsilon > 0$, there exists a positive probability path from each vertex z' to the vertex z. Hence there exists at least one z-tree T^* for which the sum of the resistances is finite, hence $\gamma(z)$ is finite. We are going to show that the stochastically stable states minimize $\gamma(z)$. Then we shall show that γ is constant on each recurrence class E_j, and that its value is γ_j—the stochastic potential of class j defined above.

LEMMA 3.2. *Let P^ε be a regular perturbed Markov process, and for each $\varepsilon \in [0, \varepsilon^*]$ let μ^ε be the unique stationary distribution of P^ε. Then $\lim_{\varepsilon \to 0} \mu^\varepsilon = \mu^0$ exists, μ^0 is a stationary distribution of P^0, and $\mu^0(z) > 0$ if and only if γ achieves its minimum at z.*

PROOF. For each state $z \in Z$ let

$$\nu^{\varepsilon}(z) = \sum_{T^* \in \mathcal{T}_z^*} \prod_{(z',z'') \in T^*} P^{\varepsilon}_{z'z''}.$$

By lemma 3.1, the stationary distribution μ^{ε} is just the normalization of ν^{ε} that satisfies $\sum \mu^{\varepsilon}(z) = 1$. Let $\bar{\gamma} = \min_z \gamma(z)$. We are going to show that $\lim_{\varepsilon \to 0} \mu^{\varepsilon}(z) > 0$ if and only if $\gamma(z) = \bar{\gamma}$. For each z-tree $T^* \in \mathcal{T}_z^*$, the *resistance* of T^* is

$$r(T^*) = \sum_{(z',z'') \in T^*} r(z', z''). \tag{A.8}$$

Given $z \in Z$, there exists (by the definition of γ) a z-tree T^* whose resistance equals $\gamma(z)$. Consider the identity

$$\varepsilon^{-\bar{\gamma}} \prod_{(z',z'') \in T^*} P^{\varepsilon}_{z'z''} = \varepsilon^{r(T^*)-\bar{\gamma}} \prod_{(z',z'') \in T^*} \varepsilon^{-r(z',z'')} P^{\varepsilon}_{z'z''}. \tag{A.9}$$

Since $\gamma(z)$ is finite, it follows from the definition of the resistance function r that

$$0 < \lim_{\varepsilon \to 0^+} \varepsilon^{-r(z',z'')} P^{\varepsilon}_{z'z''} < \infty \text{ for every } (z', z'') \in T^*. \tag{A.10}$$

If z does not minimize γ, then $r(T^*) = \gamma(z) > \bar{\gamma}$, and it follows from (A.9) and (A.10) that

$$\lim_{\varepsilon \to 0^+} \varepsilon^{-\bar{\gamma}} \prod_{(z',z'') \in T^*} P^{\varepsilon}_{z'z''} = 0.$$

By assumption, $r(T) \geq \gamma(z) > \bar{\gamma}$ for every $T \in \mathcal{T}_z^*$, hence

$$\lim_{\varepsilon \to 0^+} \varepsilon^{-\bar{\gamma}} \nu^{\varepsilon}(z) = 0. \tag{A.11}$$

Similarly, if $r(T^*) = \gamma(z) = \bar{\gamma}$, we obtain

$$\lim_{\varepsilon \to 0^+} \varepsilon^{-\bar{\gamma}} \nu^{\varepsilon}(z) > 0. \tag{A.12}$$

From lemma 3.1 we know that

$$\mu^{\varepsilon}(z) = \varepsilon^{-\bar{\gamma}} \nu^{\varepsilon}(z) / \sum_{z \in Z} \varepsilon^{-\bar{\gamma}} \nu^{\varepsilon}(z). \tag{A.13}$$

From (A.11), (A.12), and (A.13) it follows that

$$\lim_{\varepsilon \to 0^+} \mu^\varepsilon(z) = 0 \text{ if } \gamma(z) > \bar{\gamma},$$

and

$$\lim_{\varepsilon \to 0^+} \mu^\varepsilon(z) > 0 \text{ if } \gamma(z) = \bar{\gamma}.$$

In particular, we have shown that $\lim_{\varepsilon \to 0^+} \mu^\varepsilon = \mu^0$ exists, and that its support is precisely the set of states z that minimize $\gamma(z)$. Since μ^ε satisfies the equation $\mu^\varepsilon P^\varepsilon = \mu^\varepsilon$ for every $\varepsilon > 0$, and by assumption $P^\varepsilon \to P^0$, we conclude that $\mu^0 P^0 = \mu^0$. In other words, μ^0 is a stationary distribution of P^0. This completes the proof of lemma 3.2.

Since μ^0 is a stationary distribution of P^0, $\mu^0(z) = 0$ for every transient state z. Thus to find the stochastically stable states, it suffices to compute the function γ only on the recurrent states. We claim that in fact γ is *constant* on each recurrence class of P^0. To see this, suppose that z is in the recurrence class E_j and let T^* be a z-tree on Z with resistance $\gamma(z)$. Let y be some other vertex in E_j. Choose a path of zero resistance from z to y. The edges of this path, together with the edges of T^*, contain a y-tree T' that has the same resistance as T^*. From this it follows that $\gamma(y) \leq \gamma(z)$. A symmetric argument shows that $\gamma(y) \geq \gamma(z)$. Hence γ is constant on E_j.

To establish the theorem, it remains only to show that the value of γ on E_j is precisely the stochastic potential γ_j that was defined with respect to the graph Γ^0.

LEMMA 3.3. *For every recurrence class E_j of P^0, $\gamma(z) = \gamma_j$ for all $z \in E_j$.*

PROOF. Fix a particular state z_j in each recurrent class E_j. We shall show first that $\gamma(z_j) \leq \gamma_j$, and then we shall show the reverse inequality.

Fix a class E_j and a j-tree T_j in Γ^0 whose resistance $r(T_j)$ equals γ_j. For every $i \neq j$, there exists exactly one outgoing edge $(i, i') \in T_j$, and its weight is $r_{ii'}$. Recall that Γ^* is the complete directed graph with vertex set Z, and the weight on each edge (z, z') is the resistance $r(z, z')$. Choose a directed path in Γ^* that begins at z_i and ends at $z_{i'}$, and that has total resistance $r_{ii'}$. Call this path $\zeta^*_{ii'}$. By construction, $\zeta^*_{ii'}$ is a least resistant path from E_i to $E_{i'}$. For each $i \neq j$, choose a set of directed edges T^*_i in Γ^* that form a z_i-tree with respect to the subset of vertices

E_i. (Thus T_i^* contains $|E_i| - 1$ edges.) Since E_i is a recurrence class of P^0, the resistance of T_i^* equals zero. Finally, for each transient state $z \notin \cup E_i$, let ξ_z be a directed path from z to $\cup E_i$ having zero resistance.

Let E be the union of all of the edges in all of the trees T_i^*, $i \neq j$, together with all of the edges in the directed paths $\zeta_{ii'}^*$ such that $(i, i') \in T_j$, and all of the edges in the directed paths ξ_z for $z \notin \cup E_i$. By construction, E contains at least one directed path from every vertex in Z to the fixed vertex z_j. Therefore it contains a subset of edges that form a z_j-tree T^* in Γ^*. The resistance of T^* is clearly less than or equal to the sum of the resistances of the paths $\zeta_{ii'}^*$, so it is less than or equal to $r(T_j)$. Thus $\gamma(z_j) \leq \gamma_j = r(T_j)$ as claimed.

To show that $\gamma(z_j) \geq \gamma_j$, fix a class E_j and a least resistant z_j-tree T_j^* among all z_j-trees in the graph Γ^*. Label each of the specially chosen vertices z_i by the index "i" of the class E_i to which it belongs. These will be called *special vertices*. A *junction* in T_j^* is any vertex y with at least two incoming edges in T_j^*. If the junction y is not a special vertex, label it "i" if there exists a path of zero resistance from y to E_i. (There exists at least one such class because these are the recurrence classes of P^0; if there are several such classes choose any one of them as label.) Every labeled vertex is either a special vertex or a junction (or both), and the label identifies a recurrence class to which there is a path of zero resistance from that point. (Since a special vertex is already in a recurrence class, the path of zero resistance in this case is trivial.)

Define the *special predecessors* of a state $z \in Z$ to be the special vertices z_i that strictly precede z in the fixed tree T_j^* (i.e., such that there is a path from z_i to z in T_j^*) and such that there is no other special vertex $z_{i'}$ on the path from z_i to z. We claim that

> *If z_i is a special predecessor of the labeled vertex z, then the unique path in the tree from z_i to z has resistance at least r_{ik}, where k is the label of z.* (A.14)

Property (A.14) clearly holds for the tree T_j^* because any path from the special vertex z_i to a vertex labeled "k" can be extended by a zero resistance path to the class E_k, so the total path must have resistance at least r_{ik}. We shall now perform certain operations on the tree T_j^* that preserve property (A.14), and bring it into a form that can be mapped onto a j-tree in Γ^0. We shall do this by successively eliminating all junctions that are not special vertices.

Suppose that T_j^* contains a junction y that is not a special vertex, and

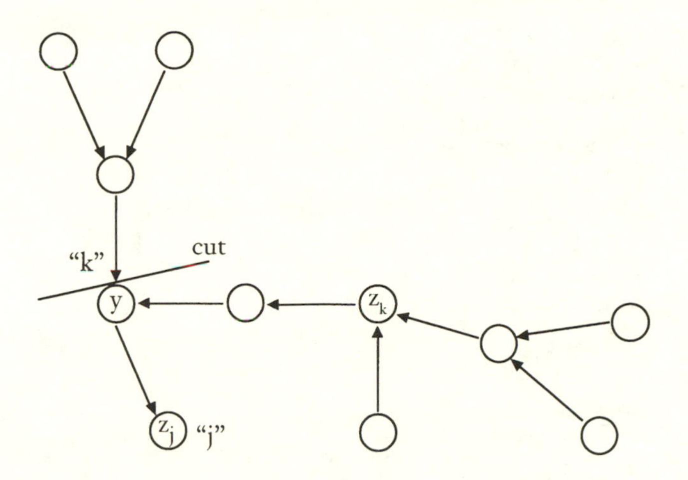

Figure A.2. Case 1 surgery: before.

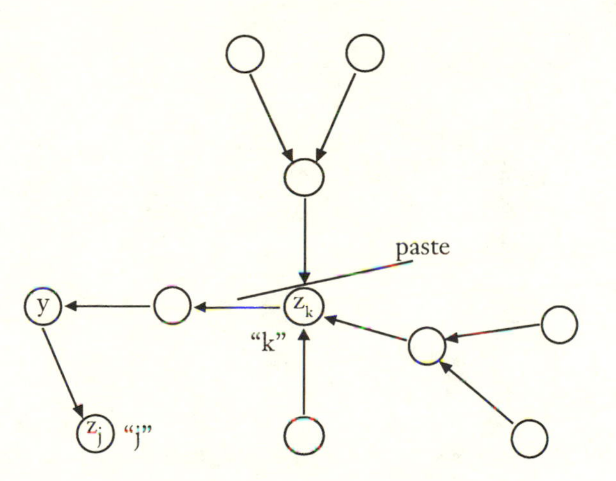

Figure A.3. Case 1 surgery: after.

let its label be "*k*." We distinguish two cases, depending on whether the special vertex z_k is or is not a predecessor of y in the tree.

CASE 1. If z_k is a predecessor of y (see Figures A.2 and A.3), cut off the subtree consisting of all edges and vertices that precede y (*except* for the path from z_k to y and all of its predecessors) and paste them onto z_k.

CASE 2. If z_k is not a predecessor of y in the tree (see Figures A.4 and

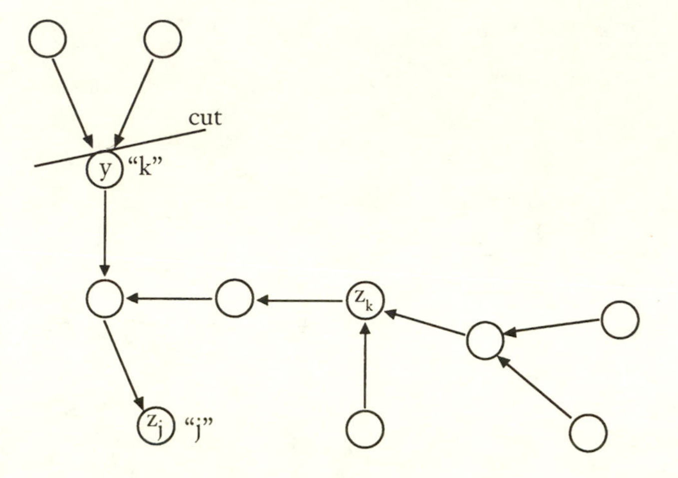

Figure A.4. Case 2 surgery: before.

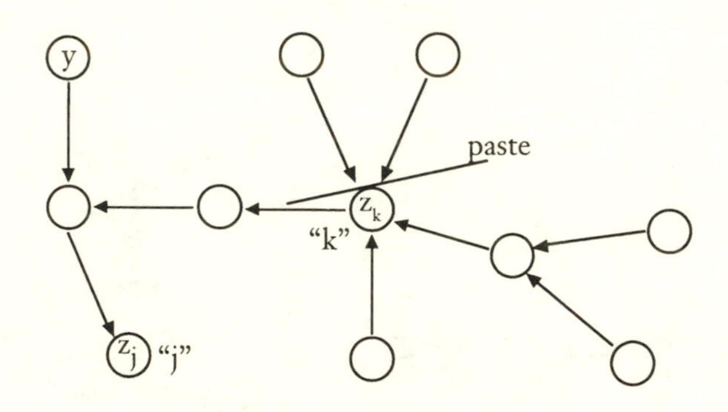

Figure A.5. Case 2 surgery: after.

A.5), cut off the subtree consisting of all edges and vertices that precede y and paste them onto the tree at the vertex z_k.

After the operation, remove y's label, since it is no longer a junction. Both of the above operations preserve property (A.14) because z_k and y had the same label. Each operation reduces by one the number of junctions that are not special vertices. Thus we eventually obtain a z_j-tree T_j^{**} in which every junction is a special vertex, property (A.14) is satisfied, and T_j^{**} has the same resistance as the original tree T_j^*.

Now construct a j-tree T^0 on the reduced graph Γ^0 as follows. For every two classes i and i' put the directed edge (i, i') in T^0 if and only if z_i is a special predecessor of $z_{i'}$ in T_j^{**}. By construction, T^0 forms a j-tree.

Let $\zeta_{ii'}^{**}$ be the unique path in T_j^{**} from z_i to $z_{i'}$. By property (A.14) its resistance is at least $r_{ii'}$. The paths $\zeta_{ii'}^{**}$ are edge-disjoint because every junction is one of the special vertices. Since T_j^{**} contains their union, the resistance of T_j^{**} is at least as large as the resistance of T^0, that is,

$$\gamma(z_j) = r(T_j^{**}) \geq \sum_{(i,i') \in T^0} r_{ii'} = r(T^0).$$

Obviously $r(T^0) \geq \gamma_j$, since the latter is the resistance of the least resistant j-tree in Γ^0. Thus $\gamma(z_j) \geq \gamma_j$ as claimed. This completes the proof of lemma 3.3, which, together with lemma 3.2 establishes theorem 3.1.

THEOREM 6.2. *Let G be a symmetric two-person coordination game with payoff matrix*

	A	B
A	a, a	c, d
B	d, c	b, b

Assume that equilibrium (A, A) is strictly risk dominant, that is, $a - d > b - c > 0$. *Let* $r^* = (b - c)/((a - d) + (b - c)) < 1/2$, *and let* $\mathcal{G}$ *be a class of graphs that are* (r, k)*–close-knit for some fixed* $r > r^*$ *and some fixed* $k \geq 1$. *Given any* $p \in (0, 1)$, *there exists a* β_p *such that for each fixed* $\beta \geq \beta_p$, *the* p*-inertia of the process* $\bar{P}^{\Gamma,\beta}$ *is bounded above for all* $\Gamma \in \mathcal{G}$, *and in particular it is bounded independently of the number of vertices in* Γ.

PROOF. Let $\mathcal{G}$ be as hypothesized in the theorem, and let Γ be a graph in $\mathcal{G}$ with vertex set V and edge set E. Choose an r–close-knit set S of size k. Consider the discrete time process in which each player in S updates according to the probability distribution in (6.3) with response rate $\beta > 0$, while each player in $\bar{S} = V - S$ always chooses action B and does not update. States of this *restricted process* $P^{\Gamma,S,\beta}$ will be denoted by $\mathbf{y}$, and states of the *unrestricted process* $P^{\Gamma,\beta}$ will be denoted by $\mathbf{x}$. Let Ξ_S denote the set of restricted states, and Ξ the set of all states.

For every two nonempty subsets of vertices $S', S'' \subseteq V$, let $E(S', S'')$ be the *set* of all undirected edges $\{i, j\}$ such that $i \in S'$ and $j \in S''$. Let $e(S', S'')$ be the *number* of such edges. The 2×2 game has the following potential function ρ:

$$\rho(A, A) = a - d \qquad \rho(A, B) = 0$$
$$\rho(B, A) = 0 \qquad \rho(B, B) = b - c$$

The corresponding potential function for the spatial game is

$$\rho^*(\mathbf{x}) = \sum_{\{i,j\}\in E} \rho(x_i, x_j).$$

(Recall that all edges are assumed to have unit weight.) It is straightforward to show (using an argument similar to that in section (6.1)) that the stationary distribution $\mu^{\Gamma,S,\beta}(\mathbf{y})$ of $P^{\Gamma,S,\beta}$ satisfies $\mu^{\Gamma,S,\beta}(\mathbf{y}) \propto e^{\beta\rho^*(\mathbf{y})}$ for all $\mathbf{y} \in \Xi_S$.

Let $\mathbf{A}_S$ denote the state in Ξ_S such that everyone in S chooses action A, and everyone in $\bar{S}$ chooses action B. We claim that $\mathbf{A}_S$ uniquely maximizes $\rho^*(\mathbf{y})$ among all restricted states $\mathbf{y}$. To see this, consider any restricted state $\mathbf{y}$ and let $S' = \{i \in S : y_i = B\}$. Substituting the values of ρ we obtain

$$\rho^*(\mathbf{y}) = (a-d)e(S-S', S-S') + (b-c)[e(S', S') + e(S', \bar{S}) + e(\bar{S}, \bar{S})],$$

and

$$\rho^*(\mathbf{A}_S) = (a-d)e(S, S) + (b-c)e(\bar{S}, \bar{S}).$$

It follows that

$$\begin{aligned}\rho^*(\mathbf{A}_S) - \rho^*(\mathbf{y}) &= (a-d)e(S', S) - (b-c)[e(S', S') + e(S', \bar{S})] \\ &= (a-d)e(S', S) - (b-c)[e(S', S') + e(S', V) - e(S', S)].\end{aligned}$$

Thus $\rho^*(\mathbf{A}_S) - \rho^*(\mathbf{y}) > 0$ if and only if

$$[(a-d) + (b-c)]e(S', S) > (b-c)[e(S', V) + e(S', S')].$$

The latter holds because by assumption

$$e(S', S)/[e(S', V) + e(S', S')] > r^* = (b-c)/[(a-d) - (b-c)].$$

(Note that $e(S', V) + e(S', S') > 0$, because there are no isolated vertices.) Thus $\mathbf{y} = \mathbf{A}_S$ uniquely maximizes $\rho^*(\mathbf{y})$ as claimed. It follows that $\mu^{\Gamma,S,\beta}$ puts arbitrarily high probability on the state $\mathbf{A}_S$ whenever β is sufficiently large.

Now fix $p \in (0, 1)$. It follows from the preceding discussion that there exists a finite value $\beta(\Gamma, S, p)$ such that $\mu^{\Gamma,S,\beta}(\mathbf{A}_S) \geq 1 - p^2/2$ for all $\beta \geq \beta(\Gamma, S, p)$. Fix such a value β. Consider the continuous time process $\bar{P}^{\Gamma,S,\beta}$ in which the agents in S update according to independent Poisson processes with unit expectation. Beginning in an initial state $\mathbf{y}^0$, let the random variable $\mathbf{y}^\tau$ denote the state of this process at time τ. As τ goes to infinity, the number of transitions of the corresponding finite process $P^{\Gamma,S,\beta}$ up to time τ also goes to infinity almost surely. Since $P^{\Gamma,S,\beta}$ is irreducible and aperiodic, it follows (see (3.10)) that $\lim_{\tau\to\infty} \Pr[\mathbf{y}^\tau = \mathbf{A}_S] = \mu^{\Gamma,S,\beta}(\mathbf{A}_S)$. Since the number of initial states is finite, there is a finite time $\tau(\Gamma, S, p, \beta)$ such that, from any initial state $\mathbf{y}^0$,

$$\forall\beta \geq \beta(\Gamma, S, p), \forall\tau \geq \tau(\Gamma, S, p, \beta), \Pr[\mathbf{y}^\tau = \mathbf{A}_S] \geq 1 - p^2.$$

Observe now that the continuous process $\bar{P}^{\Gamma,S,\beta}$ and the embedded finite process $P^{\Gamma,S,\beta}$ depend on Γ and S only through the configuration of internal edges that link vertices of S to other vertices of S, and on the configuration of external edges that link vertices of S to vertices outside of S. Since S is of size k, there are a finite number of internal edges and a finite number of ways in which they can be configured. Since S is of size k and r–close-knit, there are a finite number of external edges, and a finite number of ways in which they can be configured vis-à-vis vertices outside of S. Thus for a given r and k there are a finite number of distinct processes $\bar{P}^{\Gamma,S,\beta}$ up to isomorphism. In particular, we can find $\beta(r, k, p)$ and $\tau(r, k, p, \beta)$ such that, among all graphs Γ in $\mathcal{G}$ and all r–close-knit subsets S with k vertices, the following holds independently of the initial state:

$$\forall\beta \geq \beta(r, k, p), \forall\tau \geq \tau(r, k, p, \beta), \Pr[\mathbf{y}^\tau = \mathbf{A}_S] \geq 1 - p^2. \tag{A.15}$$

For the remainder of the discussion, we shall fix r, k, and p as in the theorem. Let us also fix $\beta^* \geq \beta(r, k, p)$ and $\tau^* = \tau(r, k, p, \beta^*)$.

Let Γ be a graph in $\mathcal{G}$ with m vertices, and let S be an r–close-knit subset in Γ of size k. We denote the unrestricted process by $\bar{P}^{\Gamma,\beta^*}$ and the restricted process by $\bar{P}^{\Gamma,S,\beta^*}$. We may couple these two processes as follows. (For general results on coupled processes see Liggett (1985).) Create two disjoint isomorphic copies of the graph Γ, say Γ_1 and Γ_2, where the ith vertex in Γ_1 corresponds to the ith vertex in Γ_2. We will define a single continuous time process that mimics $\bar{P}^{\Gamma,\beta^*}$ on Γ_1, and mimics $\bar{P}^{\Gamma,S,\beta^*}$ on Γ_2. For each state $\mathbf{x}$ of the unrestricted process $\bar{P}^{\Gamma,\beta^*}$, let $q_i(A \mid \mathbf{x})$ denote the probability that i chooses A when i updates, given that the current state is $\mathbf{x}$. Similarly, for each state $\mathbf{y}$ of the restricted process $\bar{P}^{\Gamma,S,\beta^*}$, let $q_i'(A \mid \mathbf{y})$ denote the probability that i chooses A when i updates, given that the current state is $\mathbf{y}$. Note that $q_i'(A \mid \mathbf{y}) = 0$ for all $i \in \bar{S}$.

The coupled process operates as follows. The *state* at time τ is a pair $(\mathbf{x}^\tau, \mathbf{y}^\tau)$ where x_i^τ is the choice (A or B) at the ith vertex in Γ_1, and y_i^τ is the choice (A or B) at the ith vertex in Γ_2. Each matched pair of vertices in the two graphs is governed by a Poisson process with unit expectation, and these processes are independent among the m matched pairs. Thus whenever the ith agent in Γ_1 updates the ith agent in Γ_2 updates, and vice versa. Let U be a random variable that is distributed uniformly on the interval [0, 1]. Suppose that the ith pair of individuals updates at time τ. Draw a value of U at random, and denote it by u. The ith individual in Γ_1 chooses A if $u \leq q_i(A \mid \mathbf{x}^\tau)$ and chooses B if $u > q_i(A \mid \mathbf{x}^\tau)$. Similarly, the ith individual in Γ_2 chooses A if $u \leq q_i'(A \mid \mathbf{y}^\tau)$ and chooses B if $u > q_i'(A \mid \mathbf{y}^\tau)$.

For every two states $\mathbf{x}$ and $\mathbf{y}$ on Γ_1 and Γ_2 respectively, write $\mathbf{x} \geq_A \mathbf{y}$ if $y_i = A$ implies $x_i = A$ for all i. In other words, if A appears at the ith vertex in y then A appears at the ith vertex in $\mathbf{x}$. It is evident that $\mathbf{x} \geq_A \mathbf{y}$ implies $q_i(A \mid \mathbf{x}) \geq q_i'(A \mid \mathbf{y})$ for all i. Thus, if $\mathbf{x} \geq_A \mathbf{y}$ and i chooses A in the restricted process, then i chooses A in the unrestricted process. It follows that if $\mathbf{x}^\tau \geq_A \mathbf{y}^\tau$ at some time τ, then $\mathbf{x}^{\tau'} \geq_A \mathbf{y}^{\tau'}$ at all subsequent times $\tau' \geq \tau$.

Now let the coupled process begin in the initial state $\mathbf{x}^0$ on Γ_1 and $\mathbf{y}^0$ on Γ_2, where $x_i^0 = y_i^0$ for all $i \in S$, and $y_i^0 = B$ for all $i \in \bar{S}$. Obviously we have $\mathbf{x}^0 \geq_A \mathbf{y}^0$ initially, hence we have $\mathbf{x}^\tau \geq_A \mathbf{y}^\tau$ for all $\tau \geq 0$. From (A.15) and the choice of τ^* we know that

$$\forall \tau \geq \tau^*, \Pr[\mathbf{y}^\tau = \mathbf{A}_S] \geq 1 - p^2,$$

hence

$$\forall \tau \geq \tau^*, \Pr[\mathbf{x}^\tau = \mathbf{A}_S] \geq 1 - p^2.$$

This holds for every r–close-knit set S in Γ. Since every vertex i is, by hypothesis, contained in such a set, it follows that

$$\forall \tau \geq \tau^*, \Pr[x_i^\tau = A] \geq 1 - p^2.$$

Letting α^τ be the *proportion* of individuals in Γ_1 playing action A at time τ, it follows that

$$\forall \tau \geq \tau^*, E[\alpha^\tau] \geq 1 - p^2. \tag{A.16}$$

We claim that this implies

$$\forall \tau \geq \tau^*, \Pr[\alpha^\tau \geq 1 - p] \geq 1 - p. \tag{A.17}$$

If this were false, the probability would be greater than p that more than p of the individuals at time τ were playing B. But this would imply that $E[\alpha^\tau] < 1 - p^2$, contradicting (A.16). Thus (A.17) holds. It follows that the expected waiting time in $\bar{P}^{\Gamma,\beta^*}$ until at least $1-p$ of the individuals are playing action A is bounded above by $\tau^*/(1-p)$. Moreover, once such a state is reached, the probability is at least $1-p$ that the process is in such a state at all subsequent times. Since the time τ^* holds for all graphs Γ in the family $\mathcal{G}$, the p-inertia of the family of processes $\{\bar{P}^{\Gamma,\beta^*}\}_{\Gamma \in \mathcal{G}}$ is bounded as claimed. This concludes the proof of theorem 6.2.

We remark that the assumption of *strict* risk-dominance in the theorem is important. Suppose, on the contrary, that G is a 2×2 coordination game in which both equilibria are weakly risk-dominant, that is, $r^* = 1/2$. Given any graph Γ and any subset of vertices S, the number of internal edges in S divided by the sum of degrees of members of S is at most $1/2$. Hence the conditions of the theorem cannot be satisfied, because there exists no (r, k)-close-knit graph for any value of k such that $r > r^*$. Suppose, however, that we consider graphs that are exactly $(1/2, k)$-close-knit for some value of k. For example, let $\mathcal{G}$ be the family of graphs that consist of a number of disjoint, complete subgraphs (called *cliques*), each of size $k \geq 2$. These graphs are obviously $(1/2, k)$-close-knit, but the theorem fails to hold for this situation. Indeed, at any given time it is likely that about half of the cliques will be playing action A while the other half will be playing action B; moreover, the waiting time until, say, 99% of them are playing action A becomes arbitrarily large as the number of cliques goes to infinity.

THEOREM 7.1. *Let G be a weakly acyclic n-person game, and let $P^{m,s,\varepsilon}$ be adaptive play. If s/m is sufficiently small, the unperturbed process $P^{m,s,0}$ converges with probability one to a convention from any initial state. If, in addition, ε is sufficiently small, the perturbed process puts arbitrarily high probability on the convention(s) that minimize stochastic potential.*

PROOF. By theorem 3.1, it suffices to show that the recurrent classes of $P^{m,s,0}$ correspond one to one with the conventions whenever s/m is sufficiently small. Consider any initial state h^0. There is a positive probability that all n agents in the first period sample from the most recent s entries in h^0. There is also a positive probability that for the next s periods, the agents who play always draw precisely these fixed samples. (This assumes that $s/m \leq 1/2$.) Since $\varepsilon = 0$, the agents always choose best replies to their samples. Since all best replies have positive probability, there is a positive probability that the *same* strategy-tuple x^* is played for these s periods. In other words, after s periods have elapsed, the process arrives with positive probability at a state in which the last s entries equal x^*. Call this state h^1.

Since G is weakly acyclic, there exists a best-reply path $x^* = x^0, x^1, \ldots, x^k$ ending in some strict Nash equilibrium $x^k = x^{**}$. If s/m is sufficiently small, there is a positive probability that, beginning after the history h^1, x^1 will be chosen for the next s periods. Then there is a positive probability that x^2 will be chosen for the next s periods, and so forth. Continuing in this fashion, we eventually obtain a history h^k in which the last s entries consist exclusively of repetitions of $x^k = x^{**}$. (All of this assumes that $s/m \leq 1/(k+1)$.) From h^k there is a positive probability that in $m - s$ more periods the process reaches a state h^{**} consisting of m repetitions of x^{**}. Since x^{**} is a strict Nash equilibrium, h^{**} is a convention, that is, an absorbing state of $P^{m,s,0}$. We have therefore proved that there is a positive probability of reaching a convention within a number of periods that is bounded independently of the initial state. It follows that the unperturbed process converges with probability one to a convention for any s, provided that s/m is sufficiently small. This proves that the recurrent classes are precisely the conventions, and theorem 7.1 follows at once from theorem 3.1.

THEOREM 7.2. *Let G be a generic n-person game on the finite strategy space X, and let $P^{m,s,\varepsilon}$ be adaptive play. If s/m is sufficiently small and s is sufficiently large, the unperturbed process $P^{m,s,0}$ converges with probability one to a minimal curb configuration. If, in addition, ε is sufficiently small, the perturbed process puts arbitrarily high probability on the minimal curb configuration(s) that minimize stochastic potential.*

PROOF. We shall in fact prove a slightly sharper version of this result by formulating the sense in which the game must be generic. Given an n-person game G on the finite strategy space $X = \prod X_i$, let $B_i^{-1}(x_i)$ denote the set of all probability mixtures $p_{-i} \in \Delta_{-i} = \prod_{j \neq i} \Delta(X_j)$ such that x_i is a best reply to p_{-i}. We say that G is *nondegenerate in best replies*

(NDBR) if for every i and every $x_i \in X_i$, either $B_i^{-1}(x_i)$ is empty or it contains a nonempty subset that is open in the relative topology of Δ_{-i}. The set of NDBR games is open and dense in the space $R^{n|X|}$, so NDBR is a generic property.

Given a positive integer s, we say that the probability distribution $p_i \in \Delta_i$ has *precision s* if $sp_i(x_i)$ is integer for all $x_i \in X_i$. We shall denote the set of all such distributions by Δ_i^s. For each subset Y_i of X_i, let $\Delta^s(Y_i)$ denote the set of distributions $p_i \in \Delta_i^s$ such that $p_i(x_i) > 0$ implies $x_i \in Y_i$. For each positive integer s, let $BR_i^s(X_{-i})$ be the set of pure-strategy best replies by i to some product distribution $p_{-i} \in \Delta_{-i}^s(X_{-i}) = \prod_{j \neq i} \Delta_j^s(X_j)$. Similarly, $BR_i^s(Y_{-i})$ denotes the set of all best replies by i to some product distribution $p_{-i} \in \Delta_{-i}^s(Y_{-i})$.

For each product set Y and player i, define the mapping $\beta_i(Y) = Y_i \cup BR_i(Y_{-i})$, and let $\beta(Y) = \prod \beta_i(Y)$. Similarly, for each integer $s \geq 1$ let

$$\beta_i^s(Y) = Y_i \cup BR_i^s(Y_{-i}) \text{ and } \beta^s(Y) = \prod \beta_i^s(Y).$$

Clearly, $\beta^s(Y) \subseteq \beta(Y)$ for every product set Y. We claim that if G is NDBR, then $\beta^s(Y) = \beta(Y)$ for all sufficiently large s. To establish this claim, suppose that $x_i \in \beta_i(Y)$. If $x_i \in Y_i$, then obviously $x_i \in \beta_i^s(Y)$. If $x_i \in \beta_i(Y) - Y_i$, then $B_i^{-1}(x_i) \neq \emptyset$, so by hypothesis, the set of distributions $p_{-i} = \prod_{j \neq i} p_j$ to which x_i is a best reply contains a nonempty open subset of Δ_{-i}. Hence there exists an integer $s_i(x_i, Y_{-i})$ such that x_i is a best reply to some distribution $p_{-i}^s = \prod_{j \neq i} p_j^s \in \Delta_{-i}^s(Y_{-i})$ for all $s \geq s_i(x_i, Y_{-i})$. Since there are n strategy sets, all of which are finite, it follows that there is an integer $s(Y)$ such that $\beta^s(Y) = \beta(Y)$ for all $s \geq s(Y)$. Since the number of product sets Y is finite, it follows that there is an integer s^* such that $\beta^s(Y) = \beta(Y)$ for all $s \geq s^*$ and all product sets Y, as claimed.

Now consider the process $P^{m,s,0}$. We shall show that if s is large enough and s/m is small enough, the spans of the recurrent classes correspond one to one with the minimal curb sets of G. We begin by choosing s large enough that $\beta^s \equiv \beta$. Then we choose m such that $m \geq s|X|$.

Fix a recurrent class E_k of $P^{m,s,0}$, and choose any $h^0 \in E_k$ as the initial state. We shall show that the span of E_k, $S(E_k)$, is a minimal curb set. As in the proof of theorem 7.1 above, there is a positive probability of reaching a state h^1 in which the most recent s entries involve a repetition of some fixed $x^* \in X$. Note that $h^1 \in E_k$, because E_k is a recurrent class. Let $\beta^{(j)}$ denote the j-fold iteration of β and consider the nested sequence

$$\{x^*\} \subseteq \beta(\{x^*\}) \subseteq \beta^{(2)}(\{x^*\}) \subseteq \cdots \subseteq \beta^{(j)}(\{x^*\}) \subseteq \cdots. \tag{A.18}$$

Since X is finite, there exists some point at which this sequence becomes constant, say, $\beta^{(j)}(\{x^*\}) = \beta^{(j+1)}(\{x^*\}) = Y^*$. By construction, Y^* is a curb set.

Assume that the above sequence is nontrivial, that is, $\beta(\{x^*\}) \neq \{x^*\}$. (If $\beta(x^*) = \{x^*\}$, then the following argument is not needed.) There is a positive probability that, beginning after the history h^1, some $x^1 \in \beta(\{x^*\}) - \{x^*\}$ will be chosen for the next s periods. Call the resulting history h^2. Then there is a positive probability that $x^2 \in \beta(\{x^*\}) - \{x^*, x^1\}$ will be chosen for the next s periods, and so forth. Continuing in this fashion, we eventually obtain a history h^k such that all members of $\beta(\{x^*\})$, including the original x^*, appear at least s times. All we need assume is that m is large enough so that the original s repetitions of x^* have not been forgotten. This is clearly assured if $m \geq s|X|$. Continuing this argument, we see that there is a positive probability of eventually obtaining a history h^* in which all members of $Y^* = \beta^{(j)}(\{x^*\})$ appear at least s times within the last $s|Y^*|$ periods. In particular, $S(h^*)$ contains Y^*, which by construction is a curb set.

We claim that Y^* is in fact a *minimal* curb set. To establish this, let Z^* be a minimal curb set contained in Y^*, and choose $z^* \in Z^*$. Beginning with the history h^* already constructed, there is a positive probability that z^* will be chosen for the next s periods. After this, there is a positive probability that only members of $\beta(\{z^*\})$ will be chosen, or members of $\beta^{(2)}(\{z^*\})$, or members of $\beta^{(3)}(\{z^*\})$, and so on. This happens if agents always draw samples from the "new" part of the history that follows h^*, which they will do with positive probability.

The sequence $\beta^{(k)}(\{z^*\})$ eventually becomes constant with value Z^* because Z^* is a minimal curb set. Moreover, the part of the history before the s-fold repetition of x^* will be forgotten within m periods. Thus there is a positive probability of obtaining a history h^{**} such that $S(h^{**}) \subseteq Z^*$. From such a history the process $P^{m,s,0}$ can never generate a history with members that are not in Z^* because Z^* is a curb set.

Since the chain of events that led to h^{**} began with a state in E_k, which is a recurrent class, h^{**} is also in E_k; moreover, every state in E_k is reachable from h^{**}. It follows that $Y^* \subseteq S(E_k) \subseteq Z^*$, from which we conclude that $Y^* = S(E_k) = Z^*$ as claimed.

Conversely, we must show that *if* Y^* is a minimal curb set, then $Y^* = S(E_k)$ for some recurrent class E_k of $P^{m,s,0}$. Choose an initial history h^0 that involves only strategies in Y^*. Starting at h^0, the process $P^{m,s,0}$ generates histories that involve no strategies that lie outside of $S(h^0)$, $\beta(S(h^0))$, $\beta^{(2)}(S(h^0))$, and so on. Since Y^* is a curb set, all of these strategies must occur in Y^*. With probability one the process eventually enters a recurrent class, say E_k. It follows that $S(E_k) \subseteq Y^*$. Since Y^* is a minimal curb set, the earlier part of the argument shows that $S(E_k) = Y^*$. This establishes the one-to-one correspondence between minimal curb sets and the recurrent classes of $P^{m,s,0}$. Theorem 7.2 now follows immediately from theorem 3.1.

REMARK. If G is degenerate in best replies, the theorem can fail. Consider the following example:

	A	B	C	D
a	$0, 1$	$0, 0$	$\sqrt{2}, 0$	$0, 0$
b	$2/(1+\sqrt{2}), 0$	$-1, 1/2$	$2/(1+\sqrt{2}), 0$	$0, 0$
c	$2, 0$	$0, 0$	$0, 1$	$0, 0$
d	$0, 0$	$1, 0$	$0, 0$	$2, 2$

In this game, c is a best reply to A, A is a best reply to a, a is a best reply to C, and C is a best reply to c. Hence any curb set that involves one or more of $\{A, C, a, c\}$ must involve all of them. Hence it must involve b, because b is a best reply to the probability mixture $1/(1+\sqrt{2})$ on A and $\sqrt{2}/(1+\sqrt{2})$ on C. Hence it must involve B, because B is a best reply to b. However, d is the unique best reply to B, and D is the unique best reply to d. From this it follows that every curb set contains (d, D). Since the latter is a minimal curb set (a strict Nash equilibrium), we conclude that $\{(d, D)\}$ is the unique minimal curb set. However, $\{(d, D)\}$ does not correspond to the unique recurrent class under adaptive learning.

To see why, consider any probability mixture (q_A, q_B, q_C, q_D) over column's strategies. It can be checked that strategy b is a best reply (by row) if and only if $q_B = 0$, $q_C = (\sqrt{2})q_A$, $q_D = 1 - (1+\sqrt{2})q_A$, and $q_A \geq 1/(2+\sqrt{2})$, in which case a and c are also best replies. In this situation at least one of q_A and q_C must be an irrational number. In adaptive learning, however, row responds to sample frequencies, which can only be rational numbers. It follows that, when $\varepsilon = 0$, row will never choose b as a best reply to sample information no matter how large s is. Thus, with finite samples, the unperturbed process has *two* recurrent classes: one with span $\{a, c\} \times \{A, C\}$, the other with span $\{(d, D)\}$. Hence theorem 7.2 does not hold for this (nongeneric) two person game.

THEOREM 9.1. *Let G be a two-person pure coordination game, and let $P^{m,s,\varepsilon}$ be adaptive play. If s/m is sufficiently small, every stochastically stable state is a convention; moreover, if s is sufficiently large, every such convention is efficient and approximately maximin in the sense that its welfare index is at least $(w^+ - \alpha)/(1+\alpha)$, where*

$$\alpha = w^-(1-(w^+)^2)/(1+w^-)(w^+ + w^-). \qquad \text{(A.19)}$$

PROOF. Let G be a pure coordination game with diagonal payoffs $(a_j, b_j) > (0,0)$, $1 \leq j \leq K$, and off-diagonal payoffs (0, 0). Assume that the payoffs have been normalized so that $\max_j a_j = \max_j b_j = 1$. The welfare index of the jth coordination equilibrium is defined to be $w_j = a_j \wedge b_j$, and $w^+ = \max_j w_j$. We also recall that $a^- = \min\{a_j : b_j = 1\}$, $b^- = \min\{b_j : a_j = 1\}$, and $w^- = a^- \vee b^-$.

Let $P^\varepsilon = P^{m,s,\varepsilon}$ denote adaptive learning with parameters m, s, and ε. Theorem 7.1 shows that if s/m is sufficiently small, the only recurrent classes of the unperturbed process $P^{m,s,0}$ are the K conventions $h_1, h_2, \ldots, h_K$ corresponding to the K coordination equilibria. From (9.5) we know that for every two conventions h_j and h_k, the resistance to the transition $h_j \to h_k$ is

$$r^s_{jk} = \lceil s r_{jk} \rceil \text{ where } r_{jk} = a_j/(a_j + a_k) \wedge b_j/(b_j + b_k). \qquad \text{(A.20)}$$

It will be convenient to do calculations in terms of the "reduced resistances" r_{jk}, which are more or less proportional to the actual resistances when s is large.

Construct a directed graph with K vertices, one for each of the conventions $\{h_1, h_2, \ldots, h_K\}$. Let the weight on the directed edge $h_j \to h_k$ be the reduced resistance r_{jk}. Assume that h_j is stochastically stable for some value of s. By theorem 3.1, the stochastic potential computed according to the resistances r^s_{jk} is minimized at h_j. If s is large enough, the stochastic potential computed according to the reduced resistances r_{jk} must be minimized at h_j also, that is, there is a j-tree T_j such that $r(T_j) \leq r(T)$ for all rooted trees T. (In general, $r(T)$ denotes the sum of reduced resistances of all directed edges in the set T.) We need to show that this implies

$$w_j \geq (w^+ - \alpha)/(1+\alpha).$$

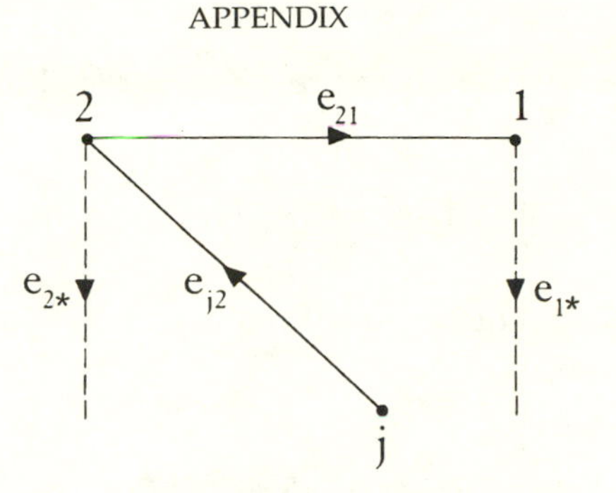

Figure A.6. The dashed edges are replaced by the solid edges to form a 1-tree.

Clearly this follows if $w_j = w^+$. Hence we may assume that $w_j < w^+$, that is, j is not a maximin equilibrium. (For brevity we shall refer to the equilibria by their indices.) Without loss of generality, let us suppose that equilibrium 1 is maximin ($w_1 = w^+$), $j \geq 2$, and in equilibrium j the row players are "worse off" than the column players: $w_j = a_j \leq b_j$.

Let j' be an equilibrium with payoffs $a_{j'} = 1$ and $b_{j'} = b^-$. Since $a_j < w^+ \leq 1$, j and j' must be distinct. Assume for the moment that $j' \neq 1$. (Later we shall dispose of the case $j' = 1$, that is, $b^- = w^+$.) Without loss of generality, let $j' = 2$: thus $a_2 = 1$ and $b_2 = b^-$.

Fix a least resistant j-tree T_j. Since $j \neq 1, 2$, there is a unique edge e_{1*} in T_j exiting from vertex 1, and a unique edge e_{2*} exiting from vertex 2 (see Figure A.6).

Delete the edges e_{1*} and e_{2*} from T_j and add the edges e_{j2} from j to 2 and e_{21} from 2 to 1. This results in a 1-tree T_1. (This is also true if e_{1*} and e_{2*} point toward the same vertex, or if e_{2*} is the same as e_{21}.) We know that $r(T_j) \leq r(T_1)$, so

$$r_{1*} + r_{2*} \leq r_{j2} + r_{21}. \tag{A.21}$$

Let us evaluate the four reduced resistances in this expression. By (A.20),

$$\begin{aligned} r_{j2} &= a_j/(a_j + a_2) \wedge b_j/(b_j + b_2) \\ &= a_j/(a_j + 1) \wedge b_j/(b_j + b^-). \end{aligned}$$

Since $w_j = a_j \leq b_j$ by hypothesis, it follows that

$$r_{j2} = w_j/(w_j + 1). \tag{A.22}$$

By (A.20), the minimum reduced resistance from 1 to any other vertex is

$$\min_k \{a_1/(a_1 + a_k) \wedge b_1/(b_1 + b_k)\}.$$

Since 1 is a maximin equilibrium, we have $a_1, b_1 \geq w^+$. Moreover, a_k, $b_k \leq 1$ for all k. Hence

$$\min_k \{a_1/(a_1 + a_k) \wedge b_1/(b_1 + b_k)\} \geq w^+/(w^+ + 1).$$

It follows that

$$r_{1^*} \geq w^+/(w^+ + 1). \qquad \text{(A.23)}$$

A similar argument shows that

$$r_{2^*} \geq b^-/(b^- + 1). \qquad \text{(A.24)}$$

Finally, since $a_1, b_1 \geq w^+$, we have

$$r_{21} = a_2/(a_2 + a_1) \wedge b_2/(b_2 + b_1) \leq 1/(1 + w^+) \wedge b^-/(b^- + w^+),$$

and in particular,

$$r_{21} \leq b^-/(b^- + w^+). \qquad \text{(A.25)}$$

Combining (A.22)–(A.25) yields the inequality

$$\frac{w^+}{w^+ + 1} + \frac{b^-}{b^- + 1} \leq \frac{w_j}{w_j + 1} + \frac{b^-}{b^- + w^+}.$$

After some algebraic manipulation, this leads to the equivalent expression

$$w_j \geq (w^+ - \alpha(b^-))/(1 + \alpha(b^-)),$$

where $\alpha(x) = [x(1 - (w^+)^2)]/[(1 + x)(w^+ + x)]$. It is straightforward to show that $\alpha(x)$ is increasing on the domain $\{x\colon x^2 \leq w^+\}$. In particular, $\alpha(b^-) \leq \alpha(w^-)$ because $b^- \leq w^- \leq w^+ \leq 1$, and therefore $(b^-)^2$, $(w^-)^2 \leq$

w^+. Letting $\alpha = \alpha(w^-)$, we therefore obtain

$$w_j \geq (w^+ - \alpha)/(1 + \alpha),$$

which is the inequality claimed in the theorem.

It remains to dispose of the case $j' = 1$, that is, the maximin equilibrium has payoffs $(1, b^-)$ and $w^+ = b^-$. Change the tree T_j by deleting the edge e_{1*} and adding the edge e_{j1}. The result is a 1-tree T_1. Now

$$r_{j1} = a_j/(a_j + a_1) \wedge b_j/(b_j + b_1) = a_j/(a_j + 1) \wedge b_j/(b_j + b^-).$$

Since $w_j = a_j \wedge b_j$ and $a_j \leq b_j$, it follows that

$$r_{j1} = a_j/(a_j + 1) = w_j/(w_j + 1).$$

On the other hand, (A.23) implies that

$$r_{1*} \geq w^+/(w^+ + 1).$$

Putting these facts together, we obtain

$$r_{1*} \geq w^+/(w^+ + 1) > w_j/(w_j + 1) = r_{j1}.$$

This implies that $r(T_1) > r(T_j)$, which contradicts the assumption that $r(T_j) \leq r(T)$ for all rooted trees T.

We now establish Pareto optimality. Suppose, by way of contradiction, that $r(T_j) \leq r(T)$ for all rooted trees T, and there exists an equilibrium $k \neq j$ that strictly dominates j:

$$a_j < a_k \text{ and } b_j < b_k. \tag{A.26}$$

Let T_j be a j-tree having minimum reduced resistance among all rooted trees. If T_j contains an edge from k to j, replace it by the oppositely directed edge e_{jk}. By (A.20) and (A.26) $r_{jk} < r_{kj}$, so this creates a k-tree whose reduced resistance is less than that of T_j, which is a contradiction. Therefore e_{kj} cannot be an edge in T_j.

Now consider any edge of form e_{hj} in T_j, where $h \neq k$. (There is at least one such edge if there is more than one coordination equilibrium.)

Replace e_{hj} with the edge e_{hk}, and make a similar substitution for all edges that point toward j. In T_j there is also a unique edge exiting from k, say, $e_{kk'}$, where $k' \neq j$ by the earlier part of the argument. Replace $e_{kk'}$ with the edge $e_{jk'}$. These substitutions yield a k-tree. We claim that it has strictly lower reduced resistance than T_j. Indeed, (A.20) and (A.26) imply that for all $h \neq j, k$, and all $k' \neq j, k$,

$$r_{hk} < r_{hj} \text{ and } r_{jk'} < r_{kk'}.$$

Hence all of the above substitutions result in a tree that has strictly lower reduced resistance. This contradiction shows that equilibrium j is not strictly Pareto dominated, and completes the proof of theorem 9.1.

NOTES

CHAPTER 1

1. For other evolutionary explanations of institutions based on game-theoretic concepts, see Schotter (1981), North (1986, 1990), Sugden (1986), and Greif (1993).

2. See, e.g., Arthur, Ermoliev, and Kaniovski, 1984; and Arthur, 1989, 1994.

3. Roughly speaking, a Markov process is *ergodic* if the limiting proportion of time spent in each state exists and is the same for almost all realizations of the process.

4. See Menger, 1871, 1883; Schotter, 1981; Marimon, McGrattan, and Sargent, 1990.

5. If the agent does not include his own previous choice in the calculation, the dynamic must be adjusted slightly near $p^t = .5$.

6. See Katz and Shapiro, 1985; Arthur, 1989.

7. Moreover, there may be decreasing unit costs resulting from economies of scale and learning-by-doing effects, which would make an early lead all the more important.

8. This discussion is based on Hopper (1982), Hamer (1986), and Lay (1992), and on personal communications by Jean-Michel Goger and Arnulf Grübler for rules of the road in France and Austria.

9. This situation was first described by Rousseau (1762) and is known in the literature as the Stag Hunt Game.

10. In biology this represents a theory of evolution, whereas here we use the term simply to describe the look of a typical dynamic path.

11. See particularly Blume, 1993, 1995a; Brock and Durlauf, 1995; Anderlini and Ianni, 1996.

CHAPTER 2

1. We shall not enter the thicket of arguments as to whether DVORAK actually is more efficient than QWERTY, though there is no question that it was *intended* to be more efficient (it was based on motion studies of typists' hand movements). For a detailed discussion of this issue see David (1985).

2. The replicator dynamic was first proposed in this form as a model of biological selection by Taylor and Jonker (1978).

3. For analytical treatments of this approach, see Weibull (1995), Binmore and Samuelson (1997), and Björnerstedt and Weibull (1996).

4. See Bush and Mosteller, 1955; Suppes and Atkinson, 1960; Arthur, 1993; Roth and Erev, 1995; Börgers and Sarin, 1995, 1996.

5. For variants of fictitious play and other belief updating processes, see Fudenberg and Kreps, 1993; Fudenberg and Levine, 1993, 1998; Crawford, 1991, 1995; Kaniovski and Young, 1995; and Benaïm and Hirsch, 1996. For experimental data on learning see, e.g., van Huyck, Battalio, and Beil, 1990, 1991; van Huyck, Battalio, and Rankin, 1995; Mookherjee and Sopher, 1994, 1997; Cheung and Friedman, 1997; and Camerer and Ho, 1997.

6. See Schotter, 1981; Kalai and Jackson, 1997.

7. Miyasawa (1961) proved that under some tie-breaking rules, every fictitious play sequence in G must converge to a Nash equilibrium of G. Monderer and Shapley (1996a) showed that under *any* tie-breaking rule, every nondegenerate game has the fictitious play property. Monderer and Sela (1996) exhibit an example of a degenerate 2×2 game and a fictitious play process that fails to converge in the sense that no limit point is a Nash equilibrium of G.

8. B^* is based on a particular tie-breaking rule. More generally, given any finite n-person game G, for each $p \in \Delta$, let $\overline{B}(p)$ be the set of all $q \in \Delta$ such that $q_i(x_i) > 0$ implies $x_i \in BR_i(p)$ for all i and all $x_i \in X_i$. The best-reply dynamic is $\dot{p}(t) = \overline{B}[p(t)] - p(t)$. This is a *differential inclusion*, that is, a set-valued differential equation. Since $\overline{B}(p)$ is upper semicontinuous, convex, and closed, the system has at least one solution from any initial point (Aubin and Cellina, 1984). For an analysis of the stability of Nash equilibrium in this process, see Hofbauer (1995).

9. Other games with the fictitious play property (under suitable tie-breaking rules) include dominance-solvable games (Milgrom and Roberts, 1990), games with strategic complementarities and diminishing returns (Krishna, 1992), and various Cournot oligopoly games (Deschamps, 1975; Thorlund-Petersen, 1990).

10. Monderer and Shapley, 1996b.

11. This idea can be substantially generalized, as we show in Chapter 7.

12. See Karni and Schmeidler (1990) for a discussion of fashion cycles. Jordan (1993) gives an example of a three-person game in which each person has two strategies and fictitious play cycles.

13. It is not essential that errors be made with uniform probability over different actions. See Chapter 5 for a discussion of this issue.

CHAPTER 3

1. This is known as the Picard-Lindelöf theorem. For a proof, see Hirsch and Smale (1974), Chapter 8, or Hartman (1982, ch. 2).

2. This and other versions of monotonicity are discussed in Nachbar (1990), Friedman (1991), Samuelson and Zhang (1992), Weibull (1995), and Hofbauer and Weibull (1996).

3. A similar result holds under even weaker forms of monotonicity (Weibull, 1995, Proposition 5.11). In the multipopulation replicator dynamics defined in (3.6), strict Nash equilibria are in fact the *only* asymptotically stable states (Hofbauer and Sigmund, 1988; Ritzberger and Vogelsberger, 1990; Weibull, 1995). If G is a symmetric two-person game played by a single population of individuals, the replicator dynamic can have asymptotically stable states that are mixed Nash equilibria. Mixed evolutionarily stable strategies (ESSs) are an example. For a discussion of this case, see Hofbauer and Sigmund (1988) and Weibull (1995).

4. For this and other standard results on finite Markov chains, see Kemeny and Snell, 1960.

5. Theorem 3.1 is analogous to results of Freidlin and Wentzell (1984) on Wiener processes. Theorem 3.1 does not follow from their results, however, because they need various regularity conditions that we do not assume; fur-

thermore, they only prove results for domains with no boundary. Analogous results for Wiener processes on bounded manifolds with reflecting boundary were obtained by Anderson and Orey (1976). The results of Anderson and Orey can be used to analyze the stochastic stability of Nash equilibria in a continuous framework, where the state space consists of the players' mixed strategies. This was the approach originally taken by Foster and Young (1990). In that paper we should have cited Anderson and Orey in addition to Freidlin and Wentzell for the construction of the potential function; this mistake was rectified in Foster and Young (1997).

6. The same conclusion holds even if everyone *strictly* prefers living in a mixed neighborhood to living in a homogeneous one. Suppose that each agent has utility u^+ if one neighbor is like himself and one is different, utility u^0 if both neighbors are like himself, and utility u^- if both neighbors are unlike himself, where $u^- < u^0 < u^+$. Then the disadvantageous trades with least resistance (lowest net utility loss) are those in which either two people of the same type trade places, or else two people of the opposite type living in mixed neighborhoods trade places. All other disadvantageous trades have a larger net utility loss and therefore a greater resistance. As in the text, this suffices to show that the segregated states have strictly lower stochastic potential than the nonsegregated states.

CHAPTER 4

1. This terminology differs slightly from Harsanyi and Selten (1988), who use the term "risk dominance" to mean *strict* risk dominance.

2. A strict Nash equilibrium is *p-dominant* if each action is the unique best reply to any belief that puts probability at least p on the other player playing his part of the equilibrium (Morris, Rob, and Shin, 1995).

3. A similar point is made by Binmore and Samuelson (1997).

CHAPTER 5

1. For discussions of urn schemes, see Eggenberger and Polya, 1923; Feller, 1950; Hill, Lane, and Sudderth, 1980; Arthur, Ermoliev, and Kaniovski, 1987.

2. The proof of theorem 5.1 is given in Kaniovski and Young (1995). Independently, Benaïm and Hirsch (1996) proved a similar result. Arthur, Ermoliev, and Kaniovski (1987) pointed out that theorem 5.1 would follow from standard arguments in stochastic approximation if the process possessed a Lyapunov function, but they did not exhibit such a function. Fudenberg and Kreps (1993) used Lyapunov theory to treat the case of 2×2 games with a unique equilibrium (which is mixed) and stochastic perturbations arising from trembles in payoffs.

CHAPTER 6

1. For related work on local interaction models and their application, see Blume (1993, 1995a, 1995b), Ellison (1993), Brock and Durlauf (1995), Durlauf (1997), Glaeser, Sacerdote, and Scheinkman (1996), and Anderlini and Ianni (1996).

2. Morris (1997) studies the amount of cohesiveness needed in an interaction

network for a strategy to be able to propagate through the network purely by diffusion.

CHAPTER 7

1. Risk dominance is only one aspect of Harsanyi and Selten's theory of equilibrium selection. In example 7.2, however, the Harsanyi-Selten theory does select the risk dominant equilibrium, that is, the equilibrium that uniquely maximizes the product of the payoffs. As shown in the text, this is not the same as the stochastically stable equilibrium. In Chapter 8 we consider a particular class of coordination games known as *Nash demand games* in which the concepts of stochastic stability and risk dominance coincide.

2. In Hurkens's model, agents sample with replacement, that is, an agent can keep resampling a single previous action by another agent without realizing that he or she looked at it before. The model also assumes that an agent can have any belief about the probability of an action as long as the agent observed it at least once. (In our model, an agent's belief is determined by the frequency of each action in the agent's sample.) These substantially different assumptions imply that the process does not discriminate between minimal curb sets even when the sample size is large: every action that occurs in some minimum curb set occurs in some stochastically stable state (Hurkens, 1995). Sanchirico (1996) considers other classes of learning models that select minimal curb sets.

3. In the same spirit, Samuelson and Zhang (1992) show that if the continuous-time replicator dynamic starts in a strictly interior state, then it eliminates (iterated) strictly dominated strategies in the sense that the proportion of the population playing such strategies approaches zero over time.

CHAPTER 8

1. Given the assumption that $a, b \leq 1/2$, it can be shown (as in the proof of Theorem 4.1) that the discrete norms are the only recurrent classes of the process $p^{\varepsilon,\delta,a,b,m}$ when $\varepsilon = 0$. When $\varepsilon > 0$ and errors are local, the process is not necessarily irreducible. Nevertheless, it has a unique recurrent class, which is the same for all $\varepsilon > 0$ and contains all the norms. The method described in the text for computing the stochastically stable norms via rooted trees remains valid in this case.

2. See Bardhan, 1984; Robertson, 1987; Winters, 1974; and Scott, 1993.

CHAPTER 10

1. The standard age of retirement in the German system, originally set at 70, was lowered during World War I to age 65, which subsequently became a standard feature of many retirement systems (Myers, 1973).

2. Catherine brought forks with her from Italy when she married the future French king, Henry II, in 1533. At that time the prevailing custom in France and elsewhere in northern Europe was to spear one's food with a sharp-pointed knife or pick it up with one's fingers (Giblin, 1987).

3. See, e.g., Mailath, Samuelson, and Shaked, 1994; Tesfatsion, 1996.

4. See Axelrod, 1997.

BIBLIOGRAPHY

Anderlini, Luc, and Antonella Ianni. 1996. "Path Dependence and Learning from Neighbors," *Games and Economic Behavior* 13:141–77.

Anderson, Robert F., and Steven Orey. 1976. "Small Random Perturbation of Dynamical Systems with Reflecting Boundary." *Nagoya Mathematics Journal* 60:189–216.

Arrow, Kenneth J. 1994. "Methodological Individualism and Social Knowledge." *American Economic Review* 84:1–9.

Arthur, W. Brian. 1989. "Competing Technologies, Increasing Returns, and Lock-In by Historical Events." *Economic Journal* 99:116–31.

______. 1993. "On Designing Economic Agents That Behave Like Human Agents." *Journal of Evolutionary Economics* 3:1–22.

______. 1994. *Increasing Returns and Path Dependence in the Economy*. Ann Arbor: University of Michigan Press.

Arthur, W. Brian, Yuri Ermoliev, and Yuri Kaniovski. 1987. "Strong Laws for a Class of Path Dependent Stochastic Processes with Applications." In *Proceedings of the International Conference on Stochastic Optimization*. Vol. 81, *Lecture Notes in Control and Information Sciences*, edited by V. Arkin, A. Shiryayev, and R. Wets. New York: Springer.

Aubin, Jean-Pierre, and A. Cellina. 1984. *Differential Inclusions*. Berlin: Springer.

Axelrod, Robert. 1984. *The Evolution of Cooperation*. New York: Basic Books.

______. 1997. "The Dissemination of Culture: A Model with Local Convergence and Global Polarization." *Journal of Conflict Resolution* 41:203–26.

Bardhan, Pranab. 1984. *Land, Labor, and Rural Poverty: Essays in Development Economics*. New York: Columbia University Press.

Basu, Kaushik, and Jörgen Weibull. 1991. "Strategy Subsets Closed under Rational Behavior." *Economics Letters* 36:141–46.

Benaïm, Michel, and Morris W. Hirsch. 1996. "Learning Processes, Mixed Equilibria, and Dynamical Systems Arising from Repeated Games." Mimeo, University of California, Berkeley.

Bergin, James, and Barton L. Lipman. 1996. "Evolution with State-Dependent Mutations." *Econometrica* 64:943–56.

Bernheim, B. Douglas. 1984. "Rationalizable Strategic Behavior." *Econometrica* 52:1007–28.

Bicchieri, Christina, Richard Jeffrey, and Brian Skyrms (eds.). 1997. *The Dynamics of Norms*. New York: Cambridge University Press.

Binmore, Ken. 1991. "Do People Exploit Their Bargaining Power: An Experimental Study." *Games and Economic Behavior* 3:295–322.

______. 1994. *Game Theory and the Social Contract*. Vol. 1, *Playing Fair*. Cambridge: MIT Press.

Binmore, Ken, and Larry Samuelson. 1994. "An Economist's Perspective on the Evolution of Norms." *Journal of Institutional and Theoretical Economics* 150:45–63.

———. 1997. "Muddling Through: Noisy Equilibrium Selection." *Journal of Economic Theory* 74:235–65.

Binmore, Ken, J. Swierzsbinski, S. Hsu, and C. Proulx. 1993. "Focal Points and Bargaining." *International Journal of Game Theory* 22:381–409.

Björnerstedt, J., and Jörgen Weibull. 1996. "Nash Equilibrium and Evolution by Imitation." In *The Rational Foundations of Economic Behavior*, edited by K. Arrow et al. London: Macmillan.

Blume, Larry. 1993. "The Statistical Mechanics of Strategic Interaction." *Games and Economic Behavior* 4:387–424.

———. 1995a. "The Statistical Mechanics of Best-Response Strategy Revision." *Games and Economic Behavior* 11:111–45.

———. 1995b. "How Noise Matters. " Mimeo, Department of Economics, Cornell University.

Börgers, Tilman, and Rajiv Sarin. 1995. "Naive Reinforcement Learning with Endogenous Aspirations." Discussion Paper, University College, London.

———. 1997. "Learning through Reinforcement and Replicator Dynamics." *Journal of Economic Theory* 77:1–14.

Brock, William A., and Steven N. Durlauf. 1995. "Discrete Choice with Social Interactions I: Theory." National Bureau of Economic Research Working Paper 5291, Cambridge, Mass.

Brown, G. W. 1951. "Iterative Solutions of Games by Fictitious Play." In *Activity Analysis of Production and Allocation*, edited by Tjalling Koopmans. New York: Wiley.

Bush, Robert, and Frederick Mosteller. 1955. *Stochastic Models for Learning*. New York: Wiley.

Camerer, Colin, and Teck-Hua Ho. 1997. "Experience-Weighted Attraction Learning in Games: A Unifying Approach." Working Paper SSW 1003, California Institute of Technology, Pasadena.

Canning, David. 1994. "Learning and Social Equilibrium in Large Populations." In *Learning and Rationality in Economics*, edited by Alan Kirman and Mark Salmon. Oxford: Blackwell.

Cheung, Yin-Wong, and Daniel Friedman. 1997. "Individual Learning in Normal Form Games." *Games and Economic Behavior* 19:46–76.

Crawford, Vincent C. 1991. "An Evolutionary Interpretation of Van Huyck, Battalio, and Beil's Experimental Results on Coordination." *Games and Economic Behavior* 3:25–59.

———. 1995. "Adaptive Dynamics in Coordination Games." *Econometrica* 63:103–44.

David, Paul A. 1985. "Clio and the Economics of QWERTY." *American Economic Review Papers and Proceedings* 75:332–37.

Deschamps, R. 1975. "An Algorithm of Game Theory Applied to the Duopoly Problem." *European Economic Review* 6:187–94.

Durlauf, Steven N. 1997. "Statistical Mechanical Approaches to Socioeconomic Behavior." In *The Economy as a Complex Evolving System*. Vol. 2, edited by W. Brian Arthur, Steven N. Durlauf, and David Lane. Redwood City, Calif.: Addison-Wesley.

Eggenberger, F., and G. Polya. 1923. "Über die Statistik verketteter Vorgänge."

Zeitschrift für Angewandte Mathematik und Mechanik 3:279–89.

Ellison, Glenn. 1993. "Learning, Local Interaction, and Coordination." *Econometrica* 61:1047–71.

______. 1995. "Basins of Attraction and Long-Run Equilibria." Department of Economics, MIT.

Ellison, Glenn, and Drew Fudenberg. 1993. "Rules of Thumb for Social Learning." *Journal of Political Economy* 101:612–43.

______. 1995. "Word of Mouth Communication and Social Learning." *Quarterly Journal of Economics* 110:93–126.

Epstein, Joshua, and Robert Axtell. 1996. *Growing Artificial Societies: Social Science from the Bottom Up*. Cambridge: MIT Press.

Feller, William. 1950. *An Introduction to Probability Theory and Its Applications*. New York: Wiley.

Foster, Dean P., and H. Peyton Young. 1990. "Stochastic Evolutionary Game Dynamics." *Theoretical Population Biology* 38:219–32.

______. 1997. "A Correction to the Paper, 'Stochastic Evolutionary Game Dynamics.'" *Theoretical Population Biology* 51:77–78.

______. 1998. "On the Nonconvergence of Fictitious Play in Coordination Games." *Games and Economic Behavior*, forthcoming.

Freidlin, Mark, and Alexander Wentzell. 1984. *Random Perturbations of Dynamical Systems*. Berlin: Springer-Verlag.

Friedman, Daniel. 1991. "Evolutionary Games in Economics." *Econometrica* 59:637–66.

Fudenberg, Drew, and Christopher Harris. 1992. "Evolutionary Dynamics with Aggregate Shocks." *Journal of Economic Theory* 57:420–41.

Fudenberg, Drew, and David Kreps. 1993. "Learning Mixed Equilibria." *Games and Economic Behavior* 5:320–67.

Fudenberg, Drew, and David Levine. 1993. "Steady State Learning and Nash Equilibrium." *Econometrica* 61:547–74.

Fudenberg, Drew, and David Levine. 1998. *The Theory of Learning in Games*. Cambridge: MIT Press.

Gaunersdorfer, Andrea, and Josef Hofbauer. 1995. "Fictitious Play, Shapley Polygons, and the Replicator Equation." *Games and Economic Behavior* 11:279–303.

Giblin, James Cross. 1987. *From Hand to Mouth: Or, How We Invented Knives, Forks, Spoons, and Chopsticks & the Tables Manners to Go with Them*. New York: Crowell.

Glaeser, Edward L., Bruce Sacerdote, and José Scheinkman. 1996. "Crime and Social Interactions." *Quarterly Journal of Economics* 111:507–48.

Greif, Avner. 1993. "Contract Enforceability and Economic Institutions in Early Trade: The Maghribi Traders' Coalition." *American Economic Review* 83:525–48.

Hamer, Mick. 1986. "Left Is Right on the Road: The History of Road Traffic Regulations." *New Scientist* 112, December 25.

Harsanyi, John, and Reinhard Selten. 1972. "A Generalized Nash Solution for Two-Person Bargaining Games with Incomplete Information." *Management Science* 18:80–106.

______. 1988. *A General Theory of Equilibrium Selection in Games*. Cambridge: MIT Press.

Hartman, Philip. 1982. *Ordinary Differential Equations*. Boston: Birkhaueser.

Hayek, von, Friedrich A. 1945. "The Use of Knowledge in Society." *American Economic Review* 35:519–30.

Hill, Bruce M., David Lane, and William Sudderth. 1980. "A Strong Law for Some Generalized Urn Processes." *Annals of Probability* 8:214–16.

Hirsch, Morris, and Steven Smale. 1974. *Differential Equations, Dynamical Systems, and Linear Algebra*. New York: Academic Press.

Hofbauer, Josef. 1995. "Stability for the Best Response Dynamics." Mimeo, University of Vienna.

Hofbauer, Josef, and Karl Sigmund. 1988. *The Theory of Evolution and Dynamical Systems*. Cambridge: Cambridge University Press.

Hofbauer, Josef, and Jörgen Weibull. 1996. "Evolutionary Selection against Dominated Strategies." *Journal of Economic Theory* 71:558–73.

Holland, John H. 1975. *Adaptation in Natural and Artificial Systems*. Ann Arbor: University of Michigan Press.

Hopper, R. H. 1982. "Left-Right: Why Driving Rules Differ." *Transportation Quarterly* 36:541–48.

Hurkens, Sjaak. 1995. "Learning by Forgetful Players." *Games and Economic Behavior* 11:304–29.

Jackson, Matthew, and Ehud Kalai. 1997. "Social Learning in Recurring Games." *Games and Economic Behavior* 21:102–34.

Jordan, James. 1993. "Three Problems in Learning Mixed Strategy Nash Equilibria." *Games and Economic Behavior* 5:368–86.

Kandori, Michihiro, George Mailath, and Rafael Rob. 1993. "Learning, Mutation, and Long-Run Equilibria in Games." *Econometrica* 61:29–56.

Kandori, Michihiro, and Rafael Rob. 1995. "Evolution of Equilibria in the Long Run: A General Theory and Applications." *Journal of Economic Theory* 65:29–56.

Kaniovski, Yuri, and H. Peyton Young. 1995. "Learning Dynamics in Games with Stochastic Perturbations." *Games and Economic Behavior* 11:330–63.

Karlin, Samuel, and H. M. Taylor. 1975. *A First Course in Stochastic Processes*. New York: Academic Press.

Karni, Edi, and David Schmeidler. 1990. "Fixed Preferences and Changing Tastes." *American Economic Association Papers and Proceedings* 80:262–67.

Katz, Michael, and Carl Shapiro. 1985. "Network Externalities, Competition, and Compatibility." *American Economic Review* 75:424–40.

Kemeny, John G., and J. Laurie Snell. 1960. *Finite Markov Chains*. Princeton: Van Nostrand.

Kirman, Alan. 1993. "Ants, Rationality, and Recruitment." *Quarterly Journal of Economics* 93:137–56.

Krishna, Vijay. 1992. "Learning in Games with Strategic Complementarities." Mimeo, Harvard Business School.

Lay, Maxwell G. 1992. *Ways of the World*. New Brunswick, N.J.: Rutgers University Press.

Lewis, David. 1969. *Convention: A Philosophical Study*. Cambridge: Harvard University Press.

Liggett, Thomas M. 1985. *Interacting Particle Systems*. New York: Springer-Verlag.

Mailath, George, Larry Samuelson, and Avner Shaked. 1994. "Evolution and Endogenous Interactions." Mimeo, Department of Economics, University of Pennsylvania, Philadelphia.

Marimon, Ramon, Ellen McGrattan, and Thomas J. Sargent. 1990. "Money as a Medium of Exchange in an Economy with Artificially Intelligent Agents." *Journal of Economic Dynamics and Control* 14:329–73.

Matsui, A. 1992. "Best Response Dynamics and Socially Stable Strategies." *Journal of Economic Theory* 57:343–62.

Maynard Smith, John. 1982. *Evolution and the Theory of Games*. Cambridge: Cambridge University Press.

Maynard Smith, John, and G. R. Price. 1973. "The Logic of Animal Conflict." *Nature* 246:15–18.

Menger, Karl. 1871. *Grundsaetze der Volkswirtschaftslehre*. Vienna: W. Braumueller. English translation by James Dingwall and Bert F. Hoselitz, under the title *Principles of Economics*. Glencoe, Ill.: Free Press, 1950.

______. 1883. *Untersuchungen über die Methode der Sozialwissenshaften und der Politischen Oekonomie insbesondere*. Leipzig: Duncker and Humboldt. English translation by Francis J. Nock, under the title *Investigations into the Method of the Social Sciences with Special Reference to Economics*. New York: New York University Press, 1985.

Milgrom, Paul, and John Roberts. 1990. "Rationalizability, Learning and Equilibrium in Games with Strategic Complementarities." *Econometrica* 58:1255–77.

Miyasawa, K. 1961. "On the Convergence of the Learning Process in a 2×2 Non-Zero-Sum Two Person Game." Research Memorandum No. 33, Economic Research Program, Princeton University.

Monderer, Dov, and A. Sela. 1996. "A 2×2 Game without the Fictitious Play Property." *Games and Economic Behavior* 14:144–48.

Monderer, Dov, and Lloyd Shapley. 1996a. "Fictitious Play Property for Games with Identical Interests." *Journal of Economic Theory* 68:258–65.

______. 1996b. "Potential Games." *Games and Economic Behavior* 14:124–43.

Mookherjee, Dilip, and Barry Sopher. 1994. "Learning Behavior in an Experimental Matching Pennies Game." *Games and Economic Behavior* 7:62–91.

______. 1997. "Learning and Decision Costs in Experimental Constant Sum Games." *Games and Economic Behavior* 19:97–132.

Morris, Stephen. 1997. "Contagion." Mimeo, Department of Economics, University of Pennsylvania.

Myers, Robert J. 1973. "Bismark and the Retirement Age." *The Actuary*, April.

Myerson, Roger B., Gregory B. Pollack, and Jeroen M. Swinkels. 1990. "Viscous Population Equilibria." *Games and Economic Behavior* 3:101–9.

Nachbar, John. 1990. "'Evolutionary' Selection Dynamics in Games: Convergence and Limit Properties." *International Journal of Game Theory* 19:59–89.

Nash, John. 1950. "The Bargaining Problem." *Econometrica* 18:155–62.

Nelson, Richard, and Sydney Winter. 1982. *An Evolutionary Theory of Economic Change*. Cambridge: Harvard University Press.

Nöldeke, Georg, and Larry Samuelson. 1993. "An Evolutionary Analysis of

Backward and Forward Induction." *Games and Economic Behavior* 5:425–54.
North, Douglass C. 1981. *Structure and Change in Economic History.* New York: Norton.
———. 1990. *Institutions, Institutional Change, and Economic Performance.* New York: Cambridge University Press.
Nydegger, R. V., and Guillermo Owen. 1974. "Two-Person Bargaining: An Experimental Test of the Nash Axioms." *International Journal of Game Theory* 3:239–49.
Pearce, David. 1984. "Rationalizable Strategic Behavior and the Problem of Perfection." *Econometrica* 52:1029–50.
Rawls, John. 1971. *A Theory of Justice.* Cambridge: Harvard University Press.
Ritzberger, Klaus, and Klaus Vogelsberger. 1990. "The Nash Field." IAS Research Report No. 263, Vienna.
Ritzberger, Klaus, and Jörgen Weibull. 1995. "Evolutionary Selection in Normal-Form Games." *Econometrica* 63:1371–99.
Robertson, A. F. 1987. *The Dynamics of Productive Relationships.* Cambridge: Cambridge University Press.
Robinson, Julia. 1951. "An Iterative Method of Solving a Game." *Annals of Mathematics* 54:296–301.
Robson, Arthur, and Fernando Vega-Redondo. 1996. "Efficient Equilibrium Selection in Evolutionary Games with Random Matching." *Journal of Economic Theory* 70:65–92.
Roth, Alvin E. 1985. "Toward a Focal Point Theory of Bargaining." In *Game-Theoretic Models of Bargaining,* edited by Alvin E. Roth. Cambridge: Cambridge University Press.
Roth, Alvin, and Ido Erev. 1995. "Learning in Extensive-Form Games: Experimental Data and Simple Dynamic Models in the Intermediate Term." *Games and Economic Behavior* 8:164–212.
Roth, Alvin E., and Michael Malouf. 1979. "Game-Theoretic Models and the Role of Information in Bargaining." *Psychological Review* 86:574–94.
Roth, Alvin E., Michael Malouf, and J. Keith Murnighan. 1981. "Sociological versus Strategic Factors in Bargaining." *Journal of Economic Behavior and Organization* 2:153–77.
Roth, Alvin E., and J. Keith Murnighan. 1982. "The Role of Information in Bargaining: An Experimental Study." *Econometrica* 50:1123–42.
Roth, Alvin, and Françoise Schoumaker. 1983. "Expectations and Reputations in Bargaining: An Experimental Study." *American Economic Review* 73:362–72.
Rousseau, Jean-Jacques. 1762. *Du contrat social, ou, principes du droit politique.* In J.-J. Rousseau, *Oeuvres complètes,* vol. 3. Dijon: Editions Gallimard, 1964.
Rubinstein, Ariel. 1982. "Perfect Equilibrium in a Bargaining Model." *Econometrica* 50:97–110.
Rumelhart, David, and James McClelland. 1986. *Parallel Distributed Processing: Explorations in the Microstructure of Cognition.* Cambridge: MIT Press.
Samuelson, Larry. 1988. "Evolutionary Foundations of Solution Concepts for Finite 2-Player, Normal-Form Games." In *Theoretical Aspects of Reasoning about Knowledge,* edited by M. Y. Vardi. Los Altos, Calif.: Morgan Kauffman.
———. 1991. "Limit Evolutionarily Stable Strategies in Two-Player, Normal

Form Games." *Games and Economic Behavior* 3:110–28.

______. 1994. "Stochastic Stability in Games with Alternative Best Replies." *Journal of Economic Theory* 64:35–65.

______. 1997. *Evolutionary Games and Equilibrium Selection*. Cambridge: MIT Press.

Samuelson, Larry, and J. Zhang, 1992. "Evolutionary Stability in Asymmetric Games." *Journal of Economic Theory* 57, 363–91.

Sanchirico, Chris W. 1996. "A Probabilistic Model of Learning in Games." *Econometrica* 64:1375–94.

Savage, Leonard J. 1954. *The Foundations of Statistics*. New York: Wiley.

Schelling, Thomas C. 1960. *The Strategy of Conflict*. Cambridge: Harvard University Press.

______. 1971. "Dynamic Models of Segregation." *Journal of Mathematical Sociology* 1:143–86.

______. 1978. *Micromotives and Macrobehavior*. New York: Norton.

Schotter, Andrew. 1981. *The Economic Theory of Social Institutions*. New York: Cambridge University Press.

Scott, John T. Jr. 1993. "Farm Leasing in Illinois, A State Survey." Department of Agricultural and Consumer Economics, Publication AERR-4703. University of Illinois, Urbana.

Selten, Reinhard. 1980. "A Note on Evolutionarily Stable Strategies in Asymmetric Animal Conflicts." *Journal of Theoretical Biology* 84:93–101.

______. 1991. "Evolution, Learning, and Economic Behavior." *Games and Economic Behavior* 3:3–24.

Shapley, Lloyd S. 1964. "Some Topics in Two-Person Games." In *Advances in Game Theory*, edited by Melvin Dresher, L. S. Shapley, and A. W. Tucker. Annals of Mathematics Studies No. 52. Princeton: Princeton University Press.

Skyrms, Brian. 1996. *Evolution of the Social Contract*. New York: Cambridge University Press.

Stahl, Ingolf. 1972. *Bargaining Theory*. Stockholm: Stockholm School of Economics.

Sugden, Robert. 1986. *The Evolution of Rights, Cooperation, and Welfare*. New York: Basil Blackwell.

Suppes, Patrick, and Richard Atkinson. 1960. *Markov Learning Models for Multiperson Interactions*. Stanford: Stanford University Press.

Taylor, Paul, and L. Jonker. 1978. "Evolutionary Stable Strategies and Game Dynamics." *Mathematical Biosciences* 40:145–56.

Tesfatsion, Leigh. 1995. "A Trade Network Game with Endogenous Partner Selection." In *Computational Approaches to Economic Problems*, edited by H. M. Amman, B. Rustem, and A. B. Whinston. Amsterdam: Kluwer Academic Publishers.

Thorlund-Petersen, L. 1990. "Iterative Computation of Cournot Equilibrium." *Games and Economic Behavior* 2:61–95.

Ullman-Margalit, Edna. 1977. *The Emergence of Norms*. Oxford: Oxford University Press.

van Huyck, John B., Raymond C. Battalio, and R. O. Beil. 1990. "Tacit Coordination Games, Strategic Uncertainty, and Coordination Failure." *American Economic Review* 80:234–48.

———. 1991. "Strategic Uncertainty, Equilibrium Selection, and Coordination Failure in Average Opinion Games." *Quarterly Journal of Economics* 106:885–911.

van Huyck, John B., Raymond C. Battalio, and Frederick W. Rankin. 1995. "Evidence on Learning in Coordination Games." Research Report No. 7, TAMU Economics Laboratory, Texas A&M University.

Vega-Redondo, Fernando. 1996. *Evolution, Games, and Economic Behavior*. New York: Oxford University Press.

von Neumann, John, and Oskar Morgenstern. 1944. *Theory of Games and Economic Behavior*. Princeton: Princeton University Press.

Weibull, Jörgen. 1995. *Evolutionary Game Theory*. Cambridge: MIT Press.

Winters, Donald. 1974. "Tenant Farming in Iowa: 1860–1900: A Study of the Terms of Rental Leases." *Agricultural History* 48:130–50.

Yaari, Menachem E., and Maya Bar-Hillel. 1984. "On Dividing Justly." *Social Choice and Welfare* 1:1–24.

Young, H. Peyton. 1993a. "The Evolution of Conventions." *Econometrica* 61:57–84.

———. 1993b. "An Evolutionary Model of Bargaining." *Journal of Economic Theory* 59:145–68.

———. 1995. "The Economics of Convention." *Journal of Economic Perspectives* 10:105–22.

Young, H. Peyton, and Dean P. Foster. 1991. "Cooperation in the Short and in the Long Run." *Games and Economic Behavior* 3:145–56.

INDEX

About the Author

H. Peyton Young is Scott and Barbara Black Professor of Economics at The Johns Hopkins University and Co-Director of the Center on Social and Economic Dynamics at Johns Hopkins and The Brookings Institution. He is the author of *Equity: In Theory and Practice* (Princeton).